Technology
IN THE
Ancient World

Technology
IN THE Ancient
World / Henry Hodges

with drawings by Judith Newcomer

 1972

Alfred·A·Knopf / NEW YORK

Preface

THIS BOOK is intended to be a straightforward account of the development of mankind's technology from its origins to the end of the period of Roman domination in the opening years of the fifth century A.D. It has been written specifically with the general reader in mind, and for this reason a certain amount of historical and technical paraphernalia has been abandoned. Technological terms have been reduced to the barest minimum, and absolute dates—however approximate—have been used in preference to the period system commonly used in archaeological writing. The normal habit of giving references in the text has also been foregone, since it was felt that little service would have been done by belaboring the reader with the mass of periodical literature that this type of reference would have involved. In lieu, a Bibliographical Note has been added at the end of the book; from it the chief sources of reference may be deduced.

It is quite impossible to thank adequately all those who have in one way or another helped me during the preparation of this book, especially the many colleagues with whom over the years I have been able to discuss various problems. To name each and every one would require in itself a volume: to them all I must apologize for this exiguous word of thanks. I must, however, record my indebtedness to Mrs. Judith Newcomer for her excellent illustrations; to Miss Sylvia Cookman for the enormous pains she took acquiring the other illustrations; to Mr. J. van de Watering, whose labors as designer can only be fully appreciated by those who have seen the disparate parts of the jigsaw with which he had to cope; to Mr. James Cochrane, my editor, whose advice and help at every stage in the preparation of this book have been invaluable; to my secretary, Miss Susan Johnson, for preparing the final typescript; and lastly, but in no manner least, to my wife, who had the unenviable task of preparing the first draft of the text.

Contents

Illustrations

Technology

IN THE

Ancient World

FIG. 1: *Tools and scrap metal from a cave at Heathery Burn, Durham, England; about 700 B.C.*

Frequently reconstruction of past technologies must depend on fragmentary evidence like the residue from some craftsman's workshop. This group of material, which came from the floor of a cave, contained bronze knives, pins and axes, as well as some bone tools. There was also a sword which had been broken up so that the metal could be used again, and part of a mold (see Figure 61) in which axes were cast. Debris of this sort gives some notion of how the bronzefounder worked, but leaves many questions unanswered.

1/Introduction: How We Know What We Do About Pre-literate and Early Literate Cultures and Their Environment

A NY ACCOUNT of the early development of technology is bound to resemble a rather poorly made patchwork quilt. The finished fabric will be not only oddly shaped and curiously colored but also full of holes. For this state of affairs the author is not entirely to blame, for the sources from which he can draw his information are often fragmentary, normally inadequate, and frequently misleading. It would be as well, therefore, to begin with an apology and to describe briefly the material on which the author has had to rely in order to construct this history of early technology.

The largest single source of information lies in the vast body of excavation reports written by numerous archaeologists over the past century. Sometimes one is fortunate: the excavator, by accident or by design, has tumbled upon the remains of what was once a large site devoted entirely to a single industry, and from his excavation report one gets a very clear picture of, let us say, a pottery works or an iron foundry as it was when being operated in a particular place over a limited period of time. On the whole this is a rare occurrence, and more commonly one has to rely on bits and pieces recovered during archaeological excavation. A few corroded tools, a lump or two of the raw materials from which he worked, and a handful of rejected scrap may be all that remains from some past craftsman's workshop. Even this is something: far

FIG. 2: *The grip of an iron sword from Luristan; as it appeared to the naked eye; as seen by X ray; and as it appeared when cut for examination by a metallurgist. Persia; about 800 B.C.*

Many objects from antiquity can, if given careful scientific examination, be made to reveal aspects of the techniques used in their making. This iron sword grip appeared to be made of a single piece of metal, but X-ray photography showed it to be made of many sections held in place by rivets. Later it was cut in half so that an analysis could be made of the various pieces of metal.

FIG. 3: *A group of carpenters' tools from Egypt; about 1500 B.C.*

Because of the extreme dryness of most Egyptian tombs, objects made of wood and metal have often survived there, while elsewhere the wood might have rotted away and the metal have become corroded beyond all recognition. As it is, these Egyptian carpenters' tools with copper blades and wooden handles give us a very clear picture of how the craftsmen worked. How each tool was used can be seen in the wall painting (Figure 98).

too often one is left with nothing but a blank in the archaeological record, either because the workshops one would like to know about lie in remote areas where as yet little archaeological excavation has been carried out or, as is so often the case, because the equipment used was so perishable that every trace has been lost.

Although archaeological excavation has revealed fewer workshops and fewer factories than one could wish, nevertheless nearly every object unearthed from the past can be made to tell us something about its method of manufacture, and indeed it is only because of the innumerable painstaking examinations of an enormous range of objects by archaeologists and by scientists that one can even begin to write a history of early technology. The information to be obtained from this source is extremely varied. We have hundreds, if not thousands, of chemical analyses, of materials such as glass and metals which demonstrate, in general terms, of what materials early man was making his tools and other equipment. On the other hand are reports, often running into many pages, of the examination of single objects many of which show amazing complexities. Even here, however, the record is strongly biased.

This is not a matter of dishonesty among archaeologists and scientists. It is the result of the durability and lack of durability of certain classes of material. Pottery and glass, for example, will often survive under the most extreme climatic conditions, while textiles, wood and leather are extremely prone to decay. It follows that one knows far less about the perishable materials than one does about those which are not. Equally, there are countries where conditions of burial allow even the perishable materials to survive in fairly large quantities; Egypt is undoubtedly the most important single example. One is tempted, therefore, to stretch the information to fill the gap, to suppose that what held good in ancient Egypt also held good in ancient Mesopotamia. We can remind ourselves endlessly that the environment, the raw materials, the very tradition of craftsmanship in the two areas were always different. It is indeed very frustrating to have to say that one knows nearly nothing about how the early Mesopotamian cobbler set about making a pair of shoes, whereas one has a great deal of information about the way his Egyptian counterpart worked. It is thus tempting to think of the whole of the Near East peopled with cobblers working along Egyptian lines instead of leaving gaping lacunae in our history, as in fact one should.

A third source of information lies in the pictures created by artists in antiquity. These may be wall paintings made in tombs, low reliefs cut into the walls of temples and palaces, mosaic pavements or pictures painted on pottery. Many of these illustrations are highly informative, and many, alas, are highly infuriating. When an Attic vase painter illustrated a potter at work on the side of a jug, one can be pretty sure that his drawing is right in every practical detail, but when the same artist painted a ship, for example, one is often left in considerable doubt about the accuracy of his detail. Indeed it is only too painfully clear that some vase painters had no idea either how a ship was constructed or how it was sailed. What they have given us is their impression of a ship at sea and not a blueprint from a naval architect. This fact itself is an interesting comment on the diversification of craftsmanship within that period, but it hardly helps us reconstruct early Greek shipping. Unfortunately, from our point of view, many artists were prepared only to give their impressions rather than accurate detailed drawings of things seen, and many, too, had to distort what they were drawing to fit some specific area of the surface they were decorating. Furthermore, one gets the impression that many artists were not free agents and that they had to illustrate what they were told to. As a result, for example, we have plenty of pictures of Mesopotamian monarchs hunting lions and conduct-

FIG. 4: *Transporting a colossus: part of the decoration from the walls of a tomb, Egypt; about 1800 B.C.*

A great deal of information about technological processes comes from models and illustrations—paintings, reliefs, mosaics and sculptures—made in antiquity. This illustration from an Egyptian tomb shows, for example, exactly how a colossal statue was transported: the colossus roped to the sledge; the teams of men pulling it; one man pouring water to lubricate the runners; others bringing further supplies on shoulder yokes; and even the foreman clapping his hands to give a rhythm to those doing the pulling.

FIG. 5: *Terra-cotta model of a plow team from Cyprus; about 2200 B.C.*

Not all illustrative material is as informative as the foregoing picture from Egypt. This clay model from Cyprus, for example, while clearly showing an ox-drawn plow, gives little idea of how the plow was constructed or the exact shape of the yoke. Nevertheless, it does tell us that a plow with a single handle was in use on the island about 2000 B.C.

FIG. 6: *An enigmatic scene: the impression made by a stone seal from Mesopotamia; before 2000 B.C.*

The impression has been variously described by different scholars as showing people preparing food, a banquet in progress or women making pottery. Obviously all three interpretations are possible and it would be unwise to use the scene as evidence of any one.

ing military campaigns, but for the humbler crafts there is an extraordinary blank. By contrast, one feels at times that the Egyptian rulers and aristocracy went out of their way to ensure that the people of later ages knew precisely how every craft was carried out, and that every worthwhile technology was depicted stage by stage. Because of this the historian, too, may become unwillingly and unwittingly biased, for it is difficult not to give greater emphasis to the illustrative matter from a country where the paintings and low reliefs are so informative than to a country where apparently the artists were neither allowed nor encouraged to illustrate the crafts.

Finally there are the written sources, and these are almost

FIG. 7: *A Greek drinking cup depicting two ships at sea; about 600 B.C.*

A further difficulty of interpreting pictures from antiquity lies in the fact that artists often distorted what they were illustrating. This was frequently due to the shape of the area that had to be decorated. In this example the hulls of the two ships have had to be curved to fit the surface of the bowl, while the yardarms have been drawn so as to follow the line of its rim.

more variable than any so far considered. At the one extreme lie the very early clay tablets on which the scribes kept the accounts of their masters, and at the other we have handed down to us the writings of the Greek and Roman men of science. Clay tablets were written simply as a means of keeping an account of what was what. Thus one finds records of, for example, the number of slaves employed by such and such a person, or the record of the number of chariots stored in the royal palace. There is no reason, therefore, why the early clay tablet records should be of any interest to us at all, but the fact remains that in the whole host of detail one can pick out occasionally rare pieces of information that help us to understand the technology of the period. Thus, for example, one of the scribes receives fifty measures of gold. He records this fact and sends the gold to the foundry for purification. In due course the gold is returned and now weighs only forty-seven measures. Like a good civil servant, to keep his name in the clear he records this fact. Of course, one is not told what process was involved, but at least one knows that some sort of refining was being carried out, and at the same time that the process must have been sufficiently well understood for the scribe to be able to persuade his master that everything was in order. Clearly, by the time this record was made the refining of gold was fairly well advanced as a technique. No one could say that the clay tablets are a storehouse of technological information, but nevertheless among all the other records there are occasional valuable pieces of information to be obtained.

As the ability to read and write became more widespread, so one finds more and more records that are pertinent to our subject, until there emerge at last what can best be called workshop recipes, often quite detailed, giving, for example, the formulation of a particular glass, stating not only the materials to be used but also the processes to be followed. Unfortunately even when one has these records one is not always in the clear. No doubt, to the writer, such a statement as "green stone from the Island of Birds" was perfectly intelligible as an ingredient in glassmaking, but we, alas, can identify neither the Island of Birds nor the green stone to be found on it, and we are left guessing as to what this item was in reality. One fares better when one comes to the writers of classical antiquity. Herodotus, the Greek historian and geographer, writing in the fifth century B.C., provides us with a host of interesting detail, although always, of course, as an aside to his main theme. But a lot of what he records was clearly never seen at first hand, and he is apt to prefix a statement with "They say" and from then on to recount some fantastic fairy tale. One is at pains to know how much of what Herodotus recorded was information that he acquired himself, how much was just gossip and how much travelers' tales. Even Pliny the Elder, writing in the first century B.C., is not totally reliable. His geography is a remarkable compilation, but if one were to follow some of the instructions he gives for technical processes one would finish up with nothing but a horrid mess. One is left with the impression that, while he recorded accurately as well as he could, he never really followed any technological process through in detail nor clearly understood it himself. Nevertheless, the situation is greatly simplified since the writings of classical antiquity have received a great deal of study from innumerable scholars, and it is safe to say that we understand the main content of most of what they have written and what reliance can be placed upon it. The debatable passages in their accounts and their own misunderstandings can therefore largely be discounted.

It follows that, no matter from which of the major sources our information comes, our history will be mostly a matter of reconstruction. The materials unearthed by archaeologists, the pictures, and the writings, all require interpretation. It is thus extremely difficult to write a purely objective account of the development of early technology, and the author is making no such claim. The reconstructions in this book are the author's own personal view of what things may have looked like or how processes may have been carried out, and new discoveries in the archaeological field any day could force him to change his view completely about many topics here discussed.

Before discussing the development of early technology, however, something further must be said about the background against which these events took place, for we must reconstruct not only the processes involved in the development of early technology but also the environment in which man was living at that time if we are to understand fully how each step followed the last. Most of the major technological innovations of antiquity were made within

FIG. 8

FIG. 9

DECIDUOUS AND
PINE FORESTS

MEDITERRANEAN
WOODLANDS

SWAMPS

GRASSLANDS

TEMPERATE
DESERT SHRUBS

TROPICAL
DESERT SHRUBS

SALT AND ROCK
DESERTS

MOUNTAIN
GRASSLANDS

FIG. 8: *Topography of the Near East.*

> Notice particularly the two high plateaus of Anatolia
> and Iran with their surrounding mountains, and the two
> great river systems, the Euphrates and the Nile, which
> draw their headwaters from temperate and tropical
> mountainous areas respectively.

FIG. 9: *Natural vegetation of the Near East.*

> Notice that there is far more extensive woodland than is
> present today, and also that some regions, such as
> Syria, were able to draw upon two or even three quite
> different vegetational environments.

FIG. 10: *The mineral resources of the Near East.*

> Notice the association between minerals and
> mountainous areas. The Nile and Euphrates valleys have
> been called the "Fertile Crescent"; their adjacent
> mountains can with justification be called the "Mineral
> Crescent."

the limited area of the Near East and the eastern end of the Mediterranean, and little could be more fatal than imagining that those regions were in antiquity as we know them today. Even in the past ten thousand years enormous changes have taken place which owe nothing to population changes (either migrations or explosions), nor to the recent development of cities, roads and railways. Far more fundamental is the fact that the entire ecology of the region has undergone drastic changes. What we know today as open, dusty plains or rich farmlands were, ten thousand years ago, more or less thickly forested, and within the forest lived a wide variety of wild animals. This is not to say that deserts did not exist, but rather that many hills that we know of today as barren ranges of rock were then at least lightly covered with trees, while the river valleys probably carried very dense forest cover. How the changes from this type of environment to that which we see today took place is a matter of considerable debate among scholars. Many see these alterations as a matter of climatic change resulting in a general desiccation of the area; a lowering of rainfall with the subsequent diminution of forest cover. Others see them as a result of man's activities, principally the felling of trees for fuel and building, making such inroads into the forest that eventually an insufficient number of trees were left for regeneration. Others again blame sheep and goats, the large flocks of which nibbled away at the young shoots of trees and seedlings, so preventing the regrowth of forest and causing its ultimate destruction. Clearly all three were contributory causes. While in different areas of the Near East it is probable that these three factors made their impact to varying degrees, the sum effect was identical.

If the presence of a reasonable level of tree cover was a prerequisite for the development of, for its day, an advanced technology, it was not a cause, for there are many well-wooded areas of the world where technological advance was, and still is, incredibly slow. One is tempted to ask, therefore, why the Near East was so favored. Three features of the region stand out above all others: the climate, which offers no real extremes; the morphology, which under primitive conditions allowed fairly easy transport and communications; and, probably most important of all, an enormous diversity of natural resources.

To deal with the last point first, when traveling in the Near East today, despite the arid conditions one cannot fail to be impressed by the ever-changing pattern of wide river valleys, sweeping highland plateaus, low ranges of rolling hills and towering ranges of mountains. In antiquity each would have carried its own particular flora, and the rapidly changing pattern of light and heavy forest, interspersed with savannah, meant that to many people there was available an enormous range of plant materials, as well, of course, as the animals that dwelt in each different environment. But this is only part of the story. The underlying rock, the cause as it were of the landscape, is equally varied, giving rise to a wide range of possible mineral resources. Thus in many of the hills and mountain ranges are to be found some, if not all, of the ores of the common metals, while the softer and therefore more easily carved stones so suitable for building, such as gypsum and calcite, are even more abundant. There are, too, a more limited number of rare mineral deposits to be found in specific areas, as for example soda and potash in various forms. A wealth of biological and mineral resources, however, is insufficient to account for the early development of technologies in this area, for without adequate communications probably little advance would ever have been made. Small isolated communities tend to tap only a fraction of the available resources, since a large part of technological advance depends upon the interchange of ideas, upon the borrowing of a technique or material from one technology and applying it to another.

The rivers and the sea provided both an easy means of transport and a stimulus to the development of boats and shipping, whereas before the development of water transport the river valleys themselves would have provided early man with a highway. Of the rivers of the Near East two are particularly important to us, the Euphrates and the Nile. From its source in what is today the eastern boundary of Turkey, the Euphrates runs parallel to the shore of the Black Sea for a considerable distance, within seventy miles of the coast, and then turning south passes within a hundred and twenty miles of Antioch, finally to turn southeast and meander down to the broad alluvial plain of Mesopotamia. The one river, thus, all but links the Black Sea, the Mediterranean and the Persian

Gulf, and serves as a link between three very important areas; the mountainous region of eastern Turkey, rich in mineral resources; the coastal belt of the Eastern Mediterranean; and the fertile valley of the Tigris and Euphrates. This does not mean that the Euphrates was navigable as a river throughout its entire length. Indeed, once out of the alluvial plain, the river is full of rapids and can only be used locally for river traffic. Nevertheless the river valley could have served as a pathway throughout the mountainous upper reaches, and the same is true of the Tigris and its tributaries in the Zagros Mountains and in the mountainous area of Kurdistan. Finally, the delta of these two rivers ten thousand years ago was not as we know it today. The mass of alluvium brought down by these rivers has been deposited at the delta which has thus emerged into the Persian Gulf, so that many ancient cities that stood once on the coast or on the estuary of these rivers not far from the coast are now many miles inland.

The Nile, by contrast, presents a very different picture, for its headwaters lie in equatorial Africa, climatically a very different zone from its lower reaches. Thus, although we know today that the sources of the river lay near mountainous areas rich in mineral resources, these were apparently never tapped in antiquity; and to the people living in the lower reaches of the Nile the river, instead of serving as a pathway to an area of wealth, led to a hot and sticky nowhere—a region into which the Egyptians ultimately led a number of expeditions, but not one with which trade was carried on as a normal, everyday affair. To those living in the lower reaches of the river in antiquity, neither the Sahara to the west nor the rocky Arabian Desert to the east, even if then more endowed with plant life than they are today, can have been exactly inviting, least of all suitable for agricultural pursuits. Nevertheless, both the Arabian Desert and Sinai provided the people of the Nile valley with a wide range of mineral resources, and thus they seldom had to search far afield for the materials they required, while boats and rafts on the river provided an ideal system of transport at least below and between the cataracts.

Other countries of the Eastern Mediterranean—Greece and the Greek islands, the Mediterranean coast of Anatolia and the Levant—present a rather different picture. The mountainous landscape of the Greek mainland, largely of porous limestone, allows no major river system, and what rivers there are provide only small isolated patches of alluvium on which it is possible to grow crops. Contact by overland routes was difficult. Thus we need hardly be surprised that shipping became the most important means of transport and communication. The Mediterranean climate, however, did allow the cultivation of several important crops such as the vine and the olive. These, the abundance of timber, and excellent stone for building may be said to be the only major resources of the area. Cyprus and many of the Greek islands, on the other hand, belong to a different geological formation and were endowed with a fairly wide range of minerals including, most important of all, the ores of copper; the Anatolian coast and the Levant were doubly blessed, for not only had they a Mediterranean climate and easy access to the sea but also they could draw on the wealth of their hinterlands.

So far we have spoken of ten thousand years ago almost as though it were a starting point for our history of technology, and in a manner of speaking this is true. But something must now be said about the time scale against which this history of technology must be projected. At a date that may be reckoned at roughly 10,000 B.C. the last of a long series of world-wide major climatic change had taken place. The great ice sheets that before this had covered a large part of the northern hemisphere had withdrawn to a region approximating very crudely that which they occupy today, and accompanying this withdrawal of the ice sheets were major changes in the flora. Areas peripheral to the ice sheets that had been tundra now became forested with pine. Areas to the south of this that had been covered with pine now slowly changed to deciduous forest, while tropical forests gradually spread to eliminate some of the southern areas that had once been deciduous. Until these changes had taken place or were well under way, mankind appears to have been only a hunter living by collecting fruit and berries, by fishing and by the taking of game. By about 8000 B.C. these major floral changes were largely complete—at least in the area that we are considering—and at about this point in time we begin to detect the emergence of a new mode of living. People began to explore their environment in a different way, so that ultimately there was to emerge a way of life based upon the domestication of plants and animals. As may be imagined this transition was accompanied by a considerable number of technological innovations. By the middle of the fourth millennium B.C. —that is to say at about 3500 B.C.—we find mankind taking another step which was ultimately to lead to the development of the City. We can for the moment beg the question as to what is meant by "City" and pause only to point out that early cities were not just overgrown villages. We shall see later that they were something far more complex, and that the very complexity was engendered by and gave rise to further technologies, of which the winning of copper from its ores and the making of copper alloys was one. And because bronze, the alloy of tin and copper, became the universally accepted metal from which to make tools and weapons, this period is generally referred to by archaeologists as the Bronze Age.

For approximately two thousand years the cities of the Near East grew and became more complex so that every major city increasingly extended its area of influence over an even larger region. By the middle of the second millennium B.C.—that is to say, around the year 1500 B.C. we find that the cities had grown so large and prosperous that their rulers were able to carve out what could best be termed "little empires." The introduction about the year 1000 B.C. of iron as a metal from which to make tools and weapons added some impetus to this development, but in the first few centuries of the last millennium B.C. shipping had become so advanced as a means of transport that those countries of the Eastern Mediterranean with a coastline found themselves naturally favored, for they were able to import nearly all those materials they required for their technologies. By 700 B.C. the expansion of trade had been such that we find countries minting their own monies as a means of exchange.

The shift of power to the center of the Mediterranean, specifically to Rome, can be seen as an extension of this same process, for barbarian Europe to the north, if having little to offer in the way of advanced technologies, had an enormous reserve of raw materials which could be acquired at a lower cost than from the countries of the Near East. After five centuries of Roman dom-

ination, however, the empire in Europe was overrun by barbarians; but few, if any, major technologies were lost to the ken
of man, and the world did not thereafter fall into a technological
doldrum. Byzantium, Islam, and the empires of medieval Europe
were all to make their own contributions to a pool of technical
skills that were to lay the foundations of the technologies we
know today. But the ancient world under the domination of Rome
had in fact reached a kind of climax in the technological field. By
the end of the Roman period many technologies had advanced as
far as possible with the equipment then available, and for further
progress to be made, a bigger or more complex plant was required.
Despite the fact that the Romans were quite capable of indulging
in gigantic undertakings, their technologies remained at the small-
equipment level. Thus, for example, if it was required to increase
the output of iron the number of furnaces was multiplied, but the
furnaces themselves remained the same size. Whatever the cause,
the idea of building a larger furnace and devising machinery to
work it seems to have been beyond the Roman mind. As a result,
the last few centuries of Roman domination produced very little
that was technologically new. No new raw materials were discovered, no new processes invented, and one can indeed say that
long before Rome fell all technological innovation had ceased.

It would be wrong to think that even within the limited area
of the Near East and the Eastern Mediterranean technological
development was anything like even. A new technology might
bear roots in one limited region, and then be brought to something like fruition before it spread to other areas. Even then by
some people it would be adopted immediately, while by others
it would take an incredibly long time before being accepted.
Sometimes this delay was demonstrably due to a lack of resources
in the country adopting the technology, but often it was due to
historical events about which, alas, we know very little. Thus, for
every technological development there are two important phases
and two points in time that we would like to record. First, there
is the place and time at which and in which the technology was
first developed; and second, there is the point in time at which the
same technology reached other areas beyond its cradle. But this
is a historian's ideal which, for the early periods at least, is seldom

to be achieved. The lack of precision in our methods of dating and our patchy knowledge of the past often make it impossible to say precisely where any particular development first took place, so much so that our knowledge that the technology existed at all frequently begins only when it has entered the second phase. A great deal of time can be wasted in speculating about precise origins, particularly where the evidence is meager. The results are often less credible than the deliberate fiction in the *Just So Stories* or the *Tales of Kai Lung*. If, therefore, new materials and new technologies suddenly appear out of the blue in the following chapters the reader must take it for granted not that this was how they developed but that in fact we know so little of their early origins that it is futile to attempt to record them here.

2/From Savage Hunter to Primitive Farmer: The New Skills and Tools Essential to Stock-raising and Agriculture

(?–5000 B.C.)

ONE SUSPECTS that, were we confronted by a collection of the earliest tools ever made by man, we would in all probability fail to recognize them for what they were. Such a hypothetical collection might contain a stick chewed to a rough point to be used for digging in a search for roots or grubs; a stone roughly chipped along one side to give a crude cutting edge; a tool improvised on the spur of the moment in order to cut up the carcass of some dead beast; or a club from a branch torn from a tree. We can only speculate that early man made such tools because we know that on occasion apes do so. Implements of this kind made in response to an immediate situation are usually discarded after they have been used, and they thus lack the only feature that would allow us to decide at this point in time that they were in fact tools—they lack any traditional form. Today we can only recognize the earliest tools of man when they unquestionably show a continued tradition in their making. Thus, although toolmaking hominids may well have emerged many millions of years earlier, we can only carry our history of toolmaking back two million years, while even then our knowledge is limited to one particular type of implement made of a stone that will fracture to give sharp cutting edges. Time has robbed us of the possibility of knowing whether at this remote period mankind was also making wooden tools along traditional lines.

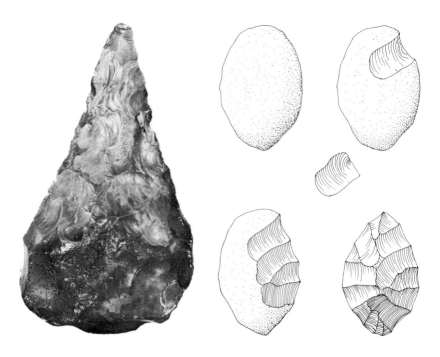

FIG. 11: *Hand ax from Gray's Inn, London; about 250,000 B.C.*

The first type of tool made by man that we can recognize as such is the so-called hand ax of flint. It was probably made for no specific purpose: it may have been used for cutting, chopping, scraping or even digging up roots. This particular implement is of great historic interest, for it was that used by John Frere in 1797 to illustrate the antiquity of man. Hitherto it had been held, as a result of taking the Bible too literally, that the world had been created in the year 4004 B.C. With John Frere and his contemporaries began the study of archaeology as we know it today, and with it the history of early technology.

FIG. 12: *Diagram showing how a hand ax was made.*

Hand axes were made either from flint or from a fine-grained rock that would provide a sharp edge. Beginning with a suitable piece of stone, flakes were hammered off in a series around the edge, so that eventually both sides of the tool were covered with scars from which the flakes had been detached. At first the flakes seem to have been discarded, but in time they were used unaltered as small tools, while later still they were often trimmed to provide knives and scrapers.

Starting with a lump or pebble of flint, or any other suitably fine-grained rock that would give a sharp edge, a series of flakes was detached by hammering with a second stone in such a way that there emerged an implement with a jagged cutting edge down each side, a crude point and a rounded, smooth butt. The function or functions of this tool, the so-called hand ax, are un-known—although in all probability it served more or less as does a pocket knife today as a tool for general and multiple uses. During the thousands of years in which this tool was in use it showed little change in form, although gradually the edges were trimmed to an even straighter line and its makers clearly aimed at producing a more geometric form. At the same time the flakes detached during the manufacture of hand axes were demonstrably used for a number of different purposes, although they were not always trimmed to any specific form.

Whatever its function, the hand ax must have proved a fairly satisfactory tool, for it remained in use over an enormous period of time. Furthermore, its distribution is curiously wide: it is found throughout Africa, a large part of Western Asia, and in Europe. Looking only at the hand axes, it is tempting to credit early man with very little intellect and, to take an extreme view, assume that the making of implements became an acquired characteristic much as in the making of nests for birds. Fortunately, however, we know that early man was beginning to master another quite different technology—the control of fire. In the caves and on the river banks where he settled a sufficiently large number of hearths have been found to show that early man at least a million years ago was building his own fires and maintaining them, although it is impossible to say how he lit them or to what purposes they were put.

At the same time that early man was learning to create neater forms of hand axes he was also beginning to pay more attention to the flakes struck off during their making, so that by some quarter of a million years ago we find groups of people who were concentrating more on the flakes struck off than on the parent block from which they were detached. The parent block was carefully trimmed in such a way that when a flake was struck off it would be of the required form and could be used directly or with the

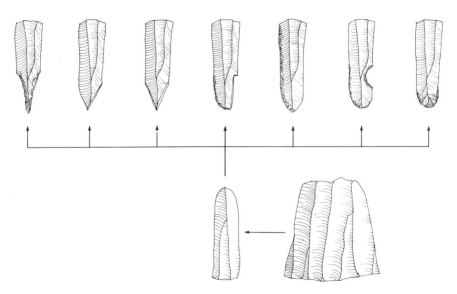

FIG. 13: *Diagram showing how stone blades were made, from about 25,000 B.C. onward.*

FIG. 14: *Diagram showing how a flint graver may have been used to carve bone.*

A new stage in the manufacture of stone tools was reached with the production of blanks from which a number of different tools could be made. The block of stone was so trimmed that from it a series of flat, rectangular pieces could be detached. By further trimming the blanks could be converted into knives, scrapers, chisels or gravers. These tools could then be used to shape other implements of wood, horn, antler and bone.

minimum amount of later shaping. These so-called flake tools were, of course, smaller than the hand axes and were thus capable of being used for finer work; it is even possible that some of them were set in a haft or handle to serve as spearheads or knife blades. It would probably be wrong to dignify these implements with the term "specialized tools," but at least a step in that direction had been taken.

It will be noticed that in these early periods, so far as man's stone tools are concerned, only a single stage of production can be detected—starting with the parent block of stone, man shaped directly the tool he required. But with the onset of the last glaciation a more sophisticated way of making stone tools was developed. Having learned how to control the shape of the flake struck from the parent block, he evolved a system of toolmaking in which long, narrow blades of flint with roughly parallel sides were struck in series from the parent core, and from this basic form a wide variety of tools could be made to meet separate purposes. One edge of a blade might be blunted so that it could serve as a small knife; the end of the blade might be rounded to form a scraper; a concavity might be chipped in the edge so that the tool served as a drawknife; or the blade might be chipped in a variety of ways to provide small chisels or gravers. Thus a new and very important stage was introduced into the making of stone tools. The parent block was no longer worked immediately into the tool, and a flake from the parent block was no longer made directly into an implement. Instead we find an intervening stage in which a "blank" was produced from which at a later stage tools could be fashioned as required.

Probably our understanding of tools not made of stone belonging to men of the last glaciation is the greater because of circumstances that allowed better preservation than before, for our knowledge of mankind of this period comes largely from the caves in the limestone rocks of southwestern France. In these caves bone and antler objects have been preserved in quantity, as well of course as the many paintings carried out on the walls. The bone and antler tools show an ever-increasing diversity. Barbed spearheads and hooks were made for the catching of fish, and bone spear throwers—long rods with a hook at one end— were used presumably for hunting larger game. From the wall paintings themselves there is sufficient evidence to suggest that the bow and arrow had been developed—although the wooden bows have long since perished—while the discovery of the occasional perforated bone needle shows that sewing or net-making was being practiced. Simple as they are, the spear thrower and the bow illustrate that mankind was at least experimenting with elementary mechanical devices.

Fires are a regular feature of caves occupied during the last Ice Age, but whether or not cooking was being practiced we do not know. We do, however, find lamps in the forms of small bowls made of soft stone, in which the fuel was probably animal fat. Indeed, some of the wall paintings were carried out in parts of the caves where daylight could not penetrate, and it has been argued that lamps had to be used in this sort of situation. The pigments used for these paintings—largely charcoal and red and yellow ocher—were pulverized in crude pestles and mortars which, like the lamps, were made of stone. There remains, how-

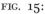

FIG. 15:
The heads of harpoons, a spear thrower, a needle and a fragment of bone from which needles have been made from various caves in southern France; about 15,000 B.C.

FIG. 16: *A modern Eskimo using a spear thrower.*

> Among the bone tools used at the end of the Ice Age
> were barbed harpoons and spear throwers. The
> harpoon was presumably set in a long wooden haft.
> The spear thrower was manipulated in much the
> same way as it still is by some peoples today, for
> example by many Eskimos. In effect it gave greater
> length to the throwing arm and thus greater range
> and power to the missile. The bone needle may have
> been used either to make nets or to sew skins together.

ever, a large field of activity during this period about which we
would like to know very much more. We have, for example, no
direct evidence that baskets were being woven, although the
presence of the bone sewing needles would suggest that the kind
of in-and-out movement required in sewing had probably also
been adapted to produce baskets and other woven objects. Furthermore, many of the bone objects are perforated, and one is
left wondering whether this work was carried out by wrist movement as, for example, when one uses a gimlet today, or whether
some type of rotary drill had already been developed.

The bone and antler tools were shaped by two principal
processes—abrasion and cutting. Blocks of sandstone and other
suitable rock were used for polishing the tools, while the cutting
was done by using flint knives and gravers, and we can infer,
therefore, the devising of tools for the making of other tools. Although this development was already implicit in earlier periods,

FIG. 17: *A stone lamp, grinding stones and pieces of various minerals used as pigments, from caves in southern France, about 15,000 B.C.*

Although man was clearly using fire a quarter of a million years ago, we know of no lamps until the end of the Ice Age. They were made by hollowing a lump of soft stone. Animal fat was probably used as a fuel and a twist of dry plant fiber as a wick. Roman lamps (Figure 235) may seem more sophisticated in shape, but their function was identical. Pigments used for decorating the walls of caves—charcoal, red and yellow ocher—were reduced to a fine powder in a simple stone mortar using a second stone as pestle. We have no way of knowing whether or not wild grain was being crushed in the same way to make flour.

as, for example, where a stone scraper had been made possibly for the manufacture of some wooden tool, we can now be perfectly certain that man was planning not one or two but often three or four stages ahead. Thus, to make a bone spearhead, man of the last glaciation would require first a suitable block of stone, shape it, from it strike off a number of blades, from these blades manufacture the tools required to shape the bone, while the spearhead itself would have to be set in a wooden haft which also had to be shaped and for which further stone tools had to be devised.

It is far from clear, however, whether any of the naturally occurring materials were being modified in other than a mechanical way. We have already noted our ignorance about cooking, but on the whole one is left with the feeling that so far mankind had not begun to experiment with chemical change. Certainly he was using no material which had been so modified that its physical characteristics were entirely altered.

The economy of man during this period was one of a hunter and a gatherer of naturally occurring wild foods, and such equip-

FIG. 18: *Bison painted on the ceiling of the cave at Lascaux, southern France; about 12,000 B.C.*

Scholars are unable to agree as to why the caves in southern France were decorated with pictures of wild animals. In this case it has been suggested that the arrowlike marks on the flanks of the bison had to do with some form of hunting magic. However this may be, two things are certain: the animals were being hunted to provide food, and the painstaking way in which the animals were depicted shows that man was not only taking interest in the shapes of things around him, but was also concerned to reproduce them accurately.

ment that has survived was devised almost entirely for the taking of game. Before we leave the Ice Age, however, we must allow ourselves one further speculation. From the period that immediately follows the Ice Age we know that the dog was already established as a distinct domesticated species, and we can therefore legitimately surmise that during the Ice Age dog and man had entered into some form of symbiosis, centered on the hunting activities of the two animals—a relationship in which the dog relied upon man's superior intellect and the man relied on the dog's superior speed, and in which the dog was rewarded with those parts of the kill unwanted by man.

The ultimate contraction of the northern ice sheet with the accompanying climatic and floral changes meant that man in Western Europe could no longer hunt his familiar game. Some species, as for example the mammoth, became extinct. Indeed, it is even possible that by the end of the Ice Age the herds of mammoth had been so drastically reduced by man's hunting activities that the species was extinct anyway, while other creatures such as the reindeer, which can survive only where its specific diet in subarctic pine forests is available, slowly migrated northward. Certainly some groups of men moved northward, too, and continued to hunt the old game, but there were others who adapted their hunting methods to the new fauna of the deciduous forest, so that we find Western Europe now peopled by small hunting groups whose game were largely red deer, wild cattle, and wild pig but who also appear to have concentrated more and more on smaller animals and birds as items in their diet. Nevertheless they remained essentially hunters and their technology continued to be that of their ancestors of the Ice Age.

We know very little about life as it was lived in the Near East during the period of the last glaciation. Few sites of this age have been examined and the conditions for preservation are not nearly as favorable as those in the caves of southwestern France, but to judge by the few stone tools that we do know, there is little reason to think that either man's way of living or his technology varied greatly from that in Western Europe. On the other hand, during the period of the last Ice Age the countries of the Near East had not undergone a period of extreme cold as had those of

Central and Western Europe. In all probability the Near East had had high rainfall and the forest cover had been that of deciduous trees, with the result that the game available during this period was largely wild cattle, wild pig and, in the more mountainous areas, the wild ancestors of our modern sheep and goats. We shall probably never know how and why the people living in this region took to the domestication of animals and to the deliberate growing of foodstuffs. For both developments numerous theories have been put forward. It seems highly probable that, long before intentional agriculture, hunting communities had deliberately protected natural stands of wild plants that they found useful in their diet, and may even have fenced them off against wild animals. It is precisely those wild species—cattle and sheep—that would have been attracted by patches of grain-growing plants that first became domesticated by man. But whatever the mechanics of the change, whatever the cause and effect, somewhere before 6000 B.C. we find a number of widely scattered communities throughout the Near East who had made the change from a pure hunting economy to that of part-time agriculturalist and stock rearer. It is possible that mankind attempted to domesticate nearly every wild species in his ken and that he succeeded with only a few, the remainder failing to breed in captivity.

The change in basic economy was accompanied by a series of technological advances, some of which may be seen directly as a result of the changed economy and others of which seem more likely to have happened as a matter of chance. Thus, in order to grow the new crops—principally wheat and barley—mankind had to clear a space in the forest. This could have been achieved by fire alone, but it would have meant that the larger timber would have been left standing, and it would seem that the ax with the polished stone blade was devised in response to the need to fell trees. Equally, while it is possible to plant grain with no more elaborate tool than a pointed stick, as indeed some primitive people still do today, a forked stick was adapted to form a primitive form of hoe.

During the period of the last glaciation the various hunting communities often seem to have created no form of habitation other than a rough shelter, and it could be argued that, because

(a)

(b)

(c)

(d)

FIG. 19: *Early nineteenth-century wood engravings by Thomas Bewick of (a) "wild" cattle from English herds kept in parks; (b) the mouflon, one of the varieties of wild sheep from which domesticated sheep were bred; (c) the wild pig; (d) the dingo, the feral dog of Australia.*

Among the animals first domesticated by man were cattle, goats, pigs and the dog. We know very little about the processes by which these animals were tamed and bred, and since we are only aware that they were domesticated by the presence of their bones found during archaeological excavations, we can seldom be certain of their exact appearance. Probably they differed very little from the wild and half-wild species illustrated by Thomas Bewick in 1807.

their game was always on the move, they, too, had to live a somewhat nomadic life, and that this state of affairs did not allow them to construct any other type of building. On the other hand, the deposits—often very deep—in many of the caves would suggest a long occupation, perhaps only seasonal, but certainly not nomadic. On the other hand, there are a few known instances, both in Central Asia and the Near East, where substantial villages had been built long before there were any manifestations of intentional farming. Of these, Jericho is undoubtedly one of the most widely known. With its excellent spring of fresh water and its dominating position over an easy crossing of the River Jordan we need not be surprised that early hunting communities chose this spot as a rendezvous. Furthermore, many primitive communities living today largely by hunting are often more sedentary than one might suppose, for the products of the chase—hides and meat—frequently require long periods for their processing, especially where simple methods of curing, either by salting or smoking, are involved. It is, of course, pure speculation that the process of smoking was one of the reasons for early man's need for fires, but the possibility should not be ignored. The fact

FIG. 20: *A wooden hoe and a flint-bladed sickle with wooden handle from Egypt; and a bone-handled sickle from Palestine. All before 4000 B.C.*

FIG. 21:
*Surviving grain from
northern Iraq; before* 5000 B.C.

FIG. 22:
*A hand mill from
Egypt; before* 4000 B.C.

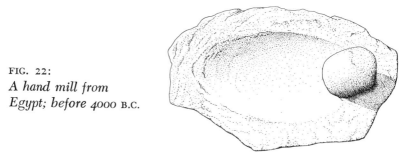

remains that small, crudely built villages clearly existed long before man put his hand to farming.

With the development of agriculture and stock breeding the technique of house building seems, nevertheless, to have developed very rapidly. In view of what has already been said about the need to clear the forest we might expect the earliest houses to have been made largely of timber, but in fact this is not so. Throughout the Near East often the earliest domestic building material was sun-dried brick, made from alluvial mud, hand-modeled and oval in section, with a strong resemblance to the loaf of bread known as a bloomar. In early villages on the Iranian Plateau and in northern Mesopotamia and in the lower levels at Jericho this type of brick was used for building, and one is tempted to suggest that its makers were more familiar with the use of rounded stones for building. To a certain extent this view is borne out by other early settlements in areas where stone of

FIG. 23: *The impression of a seal from Mesopotamia and (below) a model from Egypt showing hand mills in use; both before 2000 B.C.*

Preserved grain, mostly wheat and barley, as well as the impressions of grains left in pottery, allow us to say that these crops were being cultivated. Less can be said about early agricultural tools. Hoes made of a forked branch may well have been used at this time: they were certainly being used later in Egypt, as shown by surviving examples and by wall paintings depicting their use. Reaping was done with sickles made by setting a line of short flint blades into a bone or wooden handle. The grain was reduced to flour using a flat lower millstone and a rounded rubber. With use the lower stone became hollowed.

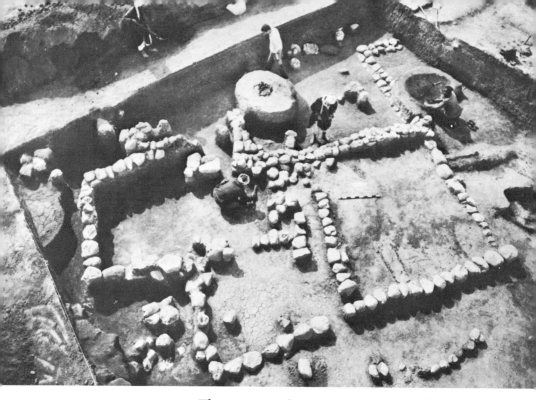

FIG. 24: *The stone foundations of huts from a village in northern Iran; before 5000 B.C.*

In mountainous areas, where rock was plentiful, houses were often built of stone and rammed earth or pisé. The photograph shows the excavation of a village at Jarmo in northern Iran. The stone footings of the houses can be seen clearly, as well as the clay-built oven (top center). It is impossible to say with certainty how high the walls were, or how the houses were roofed, when in use.

FIG. 27: *Egyptian brick molds in use, from a tomb painting about 1500 B.C. (opposite)*

Before 5000 B.C. sun-dried brick was being made in molds in Crete and western Turkey, and although one does not know the exact details, the molds were probably bottomless wooden boxes with handles for lifting. An Egyptian wall painting of a much later date shows how the process was carried out. Mud was puddled with water and packed down into the molds, which were then lifted away, leaving the newly formed brick sitting in the full sun to dry. The process is still used in exactly the same manner throughout a large part of the Near East today.

FIG. 25: *Handmade mud bricks, Jericho; before 5000 B.C.*

Even in areas where stone was plentiful it was not
always used, and throughout the Near East sun-dried
mud brick was very commonly used for building. In the
earlier stages the mud was modeled by hand, in this
case to give bricks shaped something like a bloomar
loaf. More mud was used as a mortar to join brick to
brick when building.

FIG. 26: *Diagram showing the form of single and double
brick molds used in antiquity.*

FIG. 28: *The impression of a seal showing what is probably a reed or wattle hut, Mesopotamia; about 3000* B.C.

FIG. 29: *A modern rush-matting hut from northern Nigeria.*

In the river valleys, particularly in the deltas, where reeds were readily obtained, they were probably used for building purposes, although they have long since rotted away. On seals of a much later date huts are often depicted which were clearly made from bundles of reeds lashed together, or from reed matting. Similar buildings are still being made in northern Nigeria and by the Arabs in the Euphrates delta.

suitable sizes and shapes was used for the construction of early houses. On the other hand, when we look at the earliest known settlements on the Anatolian Plateau we find a very different picture, for here, although sun-dried brick was the building material, it was evidently not hand-modeled. The bricks are too uniform and too regular to have been produced in anything other than a mold. What the molds looked like we cannot say, although many hundreds of years later we have adequate Egyptian wall paintings of bricks being made in rectangular molds; nor can we say how or where the idea developed. It is very tempting to suppose that the people using these molded bricks were initially familiar with a rock that broke naturally into a flat tabular form, but this would be pure speculation. It would be more useful to notice that before 6000 B.C. in Anatolia, at least, a simple molding device had been invented.

The people who made their houses of mud brick did not use the material solely for this purpose, for the walls of their buildings were normally plastered with a layer of mud while the floor was made of a carefully laid layer of clay. In its plastic state clay was also used for the modeling of such things as small figures and toy animals. It was also discovered during this period that, when dry, the material could be hardened by firing, and the production of pottery is often taken to be one of the more remarkable inventions made by man during this epoch.

Containers made of various materials must have been used by man even during the last Ice Age, but these were probably all made of organic materials such as hides or basketry. From a very small number of excavations where conditions of preservation have been particularly favorable it is clear that early farming communities were making vessels of wood, and it is probable that this material, too, was being used at an even earlier period. In other areas where suitably soft rocks were to be had we find early farming people making bowls of stone, while gourds and similar fruits provided mankind with a number of containers of limited shapes and dimensions. Looking only at the rather simple forms of early pottery, archaeologists have often concluded that the vessels imitate the vessels made of these different materials and that pottery to early man was essentially an ersatz material.

FIG. 30: *Pottery from Egypt; before 4000 B.C.*

Nevertheless, the earliest pottery that we know shows certain sophistications. In the first place, the clay was always mixed with some other material such as sand or crushed rock or even materials of organic origin, partly to prevent excessive shrinkage during drying and partly to make the material less prone to break during firing. Furthermore, it is seldom that we find a single type of clay mixture used for making all the pottery at any particular site, and it would seem that a great deal of attention was being paid to the properties of the finished product. Thus, we may find on the same site porous pottery that would have served well for keeping water cool and less-porous pottery that would have been more suitable for cooking vessels, while the surface of many pots was brought to a high shine by burnishing with a stone to render it even less porous. The same technique of burnishing was on occasions used equally to polish the floors and even the walls in early houses. Indeed, the technique seems to have been used for this latter purpose first and applied later to the smoothing of pottery.

The usual method of shaping vessels seems to have been, first, to mold the base of the pot over some suitable hemispherical form, often one supposes the base of an existing pot; to this were added rings of clay to extend the height of the pot—a method still widely practiced in backward parts of the world today. When dry, the pottery was probably fired either in the domestic hearth or in a bonfire especially made for the purpose, for we have no evidence of the use of kilns at this early date.

Like so many early inventions, it seems unlikely that we shall ever be able to retrace the steps by which pottery came into being and it is probable that at the back of the invention lay a long period of experimentation with vessels made of unfired clay. Nevertheless, the appearance of pottery at a time when grain-bearing plants were first being domesticated and needed to be stored, and when the grain itself had to be cooked, must be far from coincidental, while the different formulations used for the material from which the pottery was made suggest that, to its users, it was anything but an ersatz product.

The grain crops gave rise to a number of different specialized tools. One of these was the sickle, usually made by setting short lengths of flint into a suitable organic handle, and thus we find for example sickles made by removing the teeth from the lower jawbone of an animal and into the sockets of which were set short lengths of flint blade with a resin. To make foodstuff more palatable it was necessary to reduce it to a flour, and we frequently find primitive mills made from two stones—a lower flat piece of sandstone upon which the grain was placed and an upper elongated pebble of the same rock under which it was crushed. With use the lower stone became hollowed and the lower surface of the rubbing pebble became flattened. These simple mills were, of course, only an adaptation of the mortars used by man during the Ice Age to prepare his pigments. One is left wondering to what other purposes the earlier hunting communities may have put their mortars, and whether wild seeds were even then being prepared for cooking.

Sandstone rubbing stones were equally used for the polishing of ax heads and other tools made of hard fine-grained rock. At first sight it seems rather unnecessary to have bothered to bring

FIG. 31: *Men making pots on the shore of Lake Victoria.*

FIG. 32: *Men making pots in northern Nigeria.*

The discovery that clay vessels could be fired to produce the more durable pottery also belongs to this period. Methods of manufacture were simple, requiring no specialized pieces of equipment. The clay was mixed with sand or crushed rock and shaped by hand over an existing pot or was built up ring by ring. Many peoples still make their pottery in this manner today. Shapes were simple and often seem to imitate skin containers or gourds, both of which may well have been used for storage and drinking vessels long before pottery was created.

FIG. 33: *Firing pottery without a kiln, western Turkey.*

> Pottery was fired either in the domestic hearth or in special bonfires. Many people still fire their undecorated wares in this way today, and the method need not have been as unproductive as one might think. In the village in western Turkey seen in this photograph, although there are only fifty families, the women manufacture many thousands of cooking pots a year.

a stone axhead to a fine, uniform polish, but experiments carried out in Denmark and elsewhere in which trees have been felled using stone axes have shown that the polished ax was far more efficient than an axhead made simply by flaking the material roughly into shape.

Mud for the making of bricks, clay for the making of pottery and slabs of sandstone for various polishing techniques are fairly ubiquitous, but the same cannot be said for the rocks required for the making of cutting tools, and we find during this period very definite evidence of trade in these raw materials. For a people living deep in the interior or on an upland plateau or alluvial plain the fine metamorphosed rock required for making heavy tools such as axes was often to be found only in some distant mountain range, and we can only guess at the process by which it came to be brought to the village. Nevertheless, outcrops of this type of rock are fairly commonly distributed throughout the whole of the Near

FIG. 34: *Part of an Egyptian tomb painting showing stone blades being flaked; about 2000 B.C.*

FIG. 35: *A fine example of a pressure-flaked blade from Egypt; about 3500 B.C.*

Weapons, knives and other cutting tools were made of flint, obsidian or any rock that would fracture to give a sharp edge. The tools were carefully shaped, and thinned by the removal of a mass of small flakes from the surface. This was presumably done by pressing down on the object being shaped with a point made of bone or hard wood. A later Egyptian wall painting shows men trimming tools in this way.

East; but for the making of cutting tools by the older technique of flaking, the naturally occurring glass of volcanic origin, obsidian, was not only greatly favored but was also in short supply, for the number of suitable deposits of this material were very limited and recent research has shown, for example, that obsidian was being traded from a number of deposits in eastern Turkey to villages in Iraq, Mesopotamia, Palestine, Syria and western Anatolia. No matter how the trade was conducted undoubtedly it helped to open up the routes of communication between the various rather isolated village groups.

The method by which flint and obsidian were flaked during this period seems to have undergone a very radical change from that in vogue during most of the Ice Age, and there is no very obvious reason why this should be so. Tools were still made largely from the parallel-sided blade as during the last Ice Age, but the tools were considerably thinned down and made far more geometric in form by the removal of a vast number of very small flakes from the surface. This is very difficult to achieve by striking only with a stone, and it seems almost certain that these very fine flakes were removed by pressing against the edge of the flint or obsidian blade with a point of bone or hardened wood—a process often referred to as pressure-flaking. The technique was not newly dis-

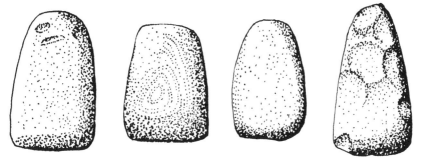

FIG. 36: *Polished stone axheads from Egypt; before 4000* B.C.

Many stone tools, especially axes, were given an over-all polish by rubbing up and down against a piece of sandstone similar to that used for making corn mills.

FIG. 37: *A complete and a fragmentary mace head from Egypt; before 4000* B.C.

FIG. 38: *Diagram showing how the bow drill was used.*

FIG. 39: *The Egyptian use of the bow drill as shown in a hieroglyph from a tomb of about 2000 B.C.*

Stone beads and sometimes axheads were perforated by means of a bow drill. The point of the drill may have been either solid or made of a section of hollow bone; sand acted as an abrasive to cut away the rock, and water was used as a lubricant. The process can, again, be seen in a later Egyptian wall painting. The broken mace head shows clearly that it was drilled first from one side and then from the other.

covered during this period for it was briefly anticipated during the last Ice Age, but as a method of manufacture it was then short-lived. By the farming communities, however, it became universally adopted throughout the whole of the Near East. Pressure-flaking is undoubtedly a knack, and a fairly difficult one to master at that. It is questionable as to whether it was adopted by every member of the community or merely by a chosen few. The use of the technique, however, allowed a far greater precision in the shaping of tools than hitherto, and thus, where before arrowheads, for example, were made by inserting a fairly irregularly shaped fragment of a blade into a shaft, we now find arrow points of greatly varying sizes, some with and some without barbs, and it looks very much as though mankind was now devoting a very great deal more thought than before to the shape and finish of his tools and weapons to make them more efficient.

So far we have spoken almost entirely of the more durable materials. For the use of wood and other plant materials we must rely upon inference and a great deal of speculation. For example, some stone tools have neatly made cylindrical holes for hafting, and these could indeed have been made by rotating a solid point of wood or bone between the palms of the handle, keeping the work lubricated with water while using sand as an abrasive. On the other hand, we know from the arrowheads and from a very limited number of wall paintings that the bow was very definitely being used by these people, and it is therefore very tempting to suppose that they had already tumbled upon the idea of the bow drill, in which the drilling point twisted into the string of the bow was rotated backward and forward.

We know that mats and baskets were being woven from such things as reeds during this period, for a very small number have survived and we occasionally have the impressions of matting left on a clay floor or on the base of a pot. It is not certain, however, when cloth textiles began to be produced, for we have no very early impressions or remains. Nevertheless, we can be pretty sure that spun thread was being manufactured. From a number of sites have come spindle whorls, the small flywheels that fit over the end of the spindle used during the making of thread. Of course, thread can be made without any equipment at all by twisting the strands

FIG. 40: *A basket from an Egyptian burial; before 4000 B.C.*

Baskets may well have been made long before the end of the Ice Age, although none have survived. By 5000 B.C. the art of basketmaking was already highly developed. Fragments of cloth show that both spinning and weaving were being practiced, but nothing has been discovered so far to show what kind of loom was in use.

FIG. 41: *Reed rafts in use in the lower Euphrates valley.*

FIG. 42: *Hollowed log canoe and raft, the Amazon River, Peru.*

The fact that people were moving from one island to another in the Mediterranean tells us that some kinds of craft were in use. In the river valleys rafts made of reed may well have been employed, but in the Mediterranean it seems more probable that substantial log rafts were used, or even canoes made by hollowing out large logs.

All three types of vessel are still in use today, as for example in Peru, where the canoes and reed boats are used mainly as personal transport, while the log raft is more commonly used for shifting heavy burdens.

of fiber between the thigh and the palm of one hand, while even a simple spindle need have no whorl, for there are primitive people today who have a spindle in which the function of the flywheel is performed by a cross of wood. One supposes, therefore, that the clay or stone spindle whorl is a fairly late development in the manufacture of thread. But it is of course one thing to make thread and quite another matter to weave a fabric from it, and thus while

(a)

(b)

(c)

FIG. 43: *The spread of farming: (a) up to 5000 B.C.;*
(b) 5000–4000 B.C.; (c) 4000–3000 B.C.

By roughly 5000 B.C. villages in which agriculture
and stock rearing were being practiced were established
throughout the Tigris-Euphrates valley, the Levant,
Anatolia, mainland Greece and on at least one island in the
eastern Mediterranean, Crete. A thousand years later, by
4000 B.C., farming had spread eastward, perhaps as far
as the Indus valley; south into the Lower Nile and North
African coastal belt; and west, up the Danube valley
into Central Europe. Within the subsequent thousand
years agriculture and animal husbandry were general
throughout Europe, and had spread throughout the
length of the Egyptian Nile and even into the larger
Sahara oases.
These maps were compiled by plotting the earliest-
known agricultural settlements in each region.
Uncertainties exist about some areas, as, for example,
the Iberian Peninsula, due to lack of archaeological
investigations that could throw light on the subject.

we may guess that occasionally textiles were being made at this
early period we have no direct evidence. Equally, since no leather
or hides have survived, and we can recognize no specific leather-
working tools, our ignorance about the use of these materials at
this period leaves an awkward gap in the record, although we may
surmise that nearly all animal remains were being used for one
purpose or another.

There is one further enormous gap in the record about which
we must say something. At Knossos, in Crete, in the lowest levels
of building, which must have been done before 5000 B.C., we
find there dwelt a people with all the aspects of early farming
that we have just outlined—a people who, incidentally, like their
neighbors in Anatolia, were molding their mud brick. These people,
therefore, must have come to the island by some form of ship or
raft. What these early seagoing craft looked like we cannot even
begin to guess, and indeed it is not until many hundreds of years
later, when we find models of boats and wall paintings, that we
have any idea of what early boats looked like, or indeed of what
they were made. Nevertheless, we are probably justified in supposing that the craft must have already been fairly developed to

allow this sort of sea passage even in fine weather, that in the rivers
and lakes of the mainland, therefore, mankind had already taken
to moving about in boats, or on rafts, and that some river transport,
even if rather limited, was being used.

By roughly 5000 B.C. early farming communities of the type
just described had been established over a somewhat limited area
of the Near East—from the more favorable valleys of mainland
Greece in the West, through southern Anatolia, in eastern Turkey,
Syria and south into Palestine, in the Mesopotamian valley and in
the upland valleys and on the plateaus of what are today northern
Iraq and Persia. As we have just seen, on some of the larger Mediterranean islands such as Cyprus and Crete similar communities
had come into being. It was from this somewhat diffuse nucleus
that the practice of farming with its accompanying technologies
spread slowly throughout Europe into Africa and northern India
during the next two thousand years or so. For the layman, often
accustomed to thinking of Egyptian civilization as having developed particularly early, it is very important to notice that by 5000
B.C., the people of the Nile were still living as primitive hunting
communities, for later we shall see the same pattern of events repeated. The idea that early Egypt achieved a higher state of technological skill than other countries of the Near East is purely
illusionary and stems from the better preservation of many of the
materials in that country than elsewhere. More commonly than
not, we shall find that Egypt received her technologies from her
neighbor countries at a rather later date than elsewhere. Communities based on the farming of a very limited number of crops and
the domestication of a few animals, with technologies based almost
entirely upon stone, clay and organic materials, are of course still
with us today. In many parts of the world they are difficult of
access, as for example in the basin of the Amazon—particularly
the upper reaches of the river—and in the highlands of New
Guinea. On the whole these people live happy and contented lives.
They have reached, as it were, a level in technological advance and
there is little spur to make them change. We shall now have to
consider why it was that, shortly after 5000 B.C., in the Near East,
we suddenly find a host of technological advances that were ultimately to affect the whole life of mankind.

3/The Spread of Farming and the Emergence of Embryonic Cities and of Writing

(5000–3000 B.C.)

I T HAS SOMETIMES been suggested that the wall paintings carried out by the hunters in southewestern France in caves during the Ice Age are unusual and represent a flowering of the arts which died with the passage of the last glaciation. This is now known to be wrong, in part, for the early farming communities that we have described in the last chapter frequently carried out quite elaborate paintings on the walls of their houses. The style is certainly different as, too, may have been the reason for carrying out these paintings, but the two groups share in common a feature that we are often apt to overlook: a certain gaiety. Material recovered from excavation is usually broken, weather-worn, and grubby and we could be forgiven for imagining that early man lived a somewhat shambling and dirty life surrounded by objects of gray and gloomy browns and reds, but the people of the Amazon and New Guinea who today live the same primitive life should warn us against this idea, for they are apt to decorate all manner of things with vivid designs carried out in the brightest colors that nature will provide. Furthermore, the early farmers were not content only to decorate their handicrafts but they appear to have decorated themselves as well, and we find small pestles and mortars used for grinding the pigments in the making of cosmetics on many of their sites. This, of course, meant that a search had to be made for suitable pigments, and this search ultimately led mankind to the ores of at least two metals. Yellow ocher or limonite, and red ocher

FIG. 44: *A cosmetic pestle and mortar from northern Iran; about 4000 B.C.*

FIG. 45: *Men preparing for a festival, Kenya. They are using green, blue and red earth pigments.*

Brightly colored minerals—red and yellow ochers and the blue and green ores of copper—were used as cosmetics. The pigments were ground to a fine powder in a small stone pestle and mortar and then probably mixed with animal fats to make rouge and eye shadow. It was, perhaps, the search for blue and green colors that first led men to deposits of metallic copper found in small quantities among the ores.

FIG. 46: *Part of an Egyptian tomb painting showing metalworkers; about 2000 B.C.*

Both gold and copper objects were at first shaped only by hammering small pieces of metal that were found to occur naturally. For this purpose rounded pebbles were used, and they continued to be employed as hammers for many centuries thereafter, as can be seen in later Egyptian tomb paintings. At first the metal was heated in the domestic hearth, but later small forges were used in which the fuel was brought to an adequate temperature by the craftsmen blowing into the fire through tubes of wood or reed.

or hematite, commonly known as jeweler's rouge, are both ores of iron, while the green mineral, malachite, and the blue mineral, azurite, are both ores of copper. But while iron never occurs as the metal in ore deposits, the same is not true of copper, and occasionally sizable lumps of metallic copper are to be found among the ores of the metal. It is therefore more than just a possibility that man's first interest in metallic copper was aroused while he was in fact searching for a suitable green pigment with which to make eye shadow.

Apart from its color and luster metallic copper had a number of properties that mankind had so far not encountered. Although it could be polished like any normal stone, using blocks of abrasive, it could not be chipped as could, for example, flint or obsidian or any of the finer-grained rocks, and attempts to chip it would show that it bent and would stay bent. It could in fact be hammered into the required shape. Thus the earliest metal artifacts that we know are small pieces of copper that have been hammered into various forms, generally rings and other small ornaments. They were small because the earliest pieces of copper available were themselves small and man had found no way of joining the fragments. Furthermore, the amount of hammering that could be used in shaping native copper was severely limited, for if hammering were carried beyond a certain point the metal became brittle and ultimately cracked, and indeed modern metallurgical examination of some of these very early metal objects shows that they had been hammered to a point very near to which they could be expected to break up. However, the stresses that build up within the metal that cause the cracking can be relieved by heating the metal in a fire to a fairly high temperature at which the metal begins to glow red. On cooling, the metal can then be hammered again to shape it until it begins to harden, when it must be heated once again to relieve the stresses that have built up. This process of relieving the stress in metal by reheating is today known as annealing, and it would seem that within a century or two of the first use of copper the idea of annealing had already been discovered. The discovery was, indeed, a very important one, chiefly because it introduced mankind to the concept that metals could be altered by the use of high temperatures. We could argue that the discovery of annealing was purely accidental—that some fragment of the metal got dropped into a fire and afterward was found to have become softened—but it could be argued equally that it was the result of a logical process. Many natural materials, as, for example, bitumen or resins, soften with heating, and it seems therefore highly probable that the use of fire to relieve the hardness of hammered copper may well have been an attempt to apply the principle of softening by heating.

The occurrence of native copper can never have been widespread, for not only is the metal rare in ore deposits, but the ores of copper themselves are very limited in occurrence. Copper ores are to be found in the mountainous areas of eastern Turkey and Syria, in the Zagros Mountains on the western edge of the Iranian Plateau, in Sinai, in the mountains of the Arabian Desert east of the Nile and of course on the island of Cyprus—from which its name is derived. Of these ore deposits those in eastern Turkey and northern Syria lie fairly and squarely in the middle of the area in which we have seen that early farming first developed and we need not be surprised, therefore, that it is precisely this area in which we find the first use of native copper for the making of small objects.

The urge to produce brightly colored things found an outlet in another technological field. The earliest pottery we know seems to have been purely utilitarian and its firing was not particularly carefully controlled. At the end of the firing process the pottery evidently lay buried or half buried in a bed of smoldering wood and ash with the result that the surface is often multicolored with scores of black and gray across a body often brown or dull red. But even in the earliest pottery we can detect an attempt on the part of its makers to do something about rectifying this rather dull appearance, and we find for example that red ocher had been burnished into the surface of the pottery before it was fired so that it would ultimately appear a uniform red. The process was not always successful, and more often than not the surface was marred by black smudges where the burning wood had fallen across the pottery and stayed there while it cooled. But at much the same time that early man was developing an interest in metallic copper he was also giving considerable thought to methods of producing pottery that would allow the colors to stand out unmarred. Essentially his problem was one of removing the pottery from the fire

FIG. 47: *Painted pottery from northern Syria; about 3500 B.C.*

Red ocher was used to paint designs on pottery, its color giving a strong contrast to a buff or white surface. The painting was clearly done with some form of brush, but it is questionable as to whether this was a brush specially made for the purpose or merely a feather dipped into the pigment.

itself, at least during the cooling stage, and initially this may have been achieved by removing the pottery from the fire on the end of a long stick and setting it down to cool in the open air—although later it was to lead to the development of a structure in which the pottery on the one hand and the fire on the other were separated. Furthermore, the potter was no longer content simply to have pots of red and his attention began to be drawn to different clays and the different colors they produced after firing. Potters now began to combine in the same vessel clays which fired white and clays which fired red under normal conditions, and we begin to find pots which are of a red body and that had been decorated on the surface with white clay or vice versa.

A few copper trinkets and a number of gaily painted pots, however, do not represent a technological advance, and their presence can have done little to alleviate the hard life of man. The importance of their appearance at this stage in man's history lies not in the end products themselves but in the fact that through them we can see that the fire was becoming no longer a domestic hearth simply for cooking, lighting and keeping wild animals at bay: it was beginning to be appreciated as a medium whereby many raw materials could be greatly changed. If we are to understand why mankind moved on from this platform of technological development we must look elsewhere for the answer.

As we have already seen, the earliest farming communities that we know of tend to lie in upland areas—in regions in fact wherein the soil is relatively light and easily worked. But these same areas have a disadvantage from the farmer's point of view in that the land may rapidly become exhausted. Thus, after a clearing had been made in the woods to establish a field, the first few crops were probably of good quality but in time the fertility of the soil fell and new clearings must have been made. This in itself restricted the possible size of any single community if the fields themselves were not to become far too distant from the village to be of practical value; apart from which hunting remained an important aspect of the economy and much of the foodstuff still had to come from wild animals, while the virgin forests in which game lived became ever distant from the habitations. In Iran, this would have meant that adjacent communities could only thrive at considerable distances from one another. It would be totally erroneous to suggest there was any shortage of land—*that* there certainly was not—but farming in the upland areas must have given rise to somewhat isolated settlements.

It would be wrong, too, to suppose that early farming people were a static population rooted to a particular piece of soil, as was the case with so many later peasant communities. Within the few centuries immediately following the year 5000 B.C., we find them spreading into Central Europe, eastward into the Indus valley and southward down into the Nile valley, and it is in the valleys of these great rivers that are to be found the greatest technological changes. The rivers themselves not only acted as roadways connecting each community with the next, but they also watered the crops and, when in spate, brought down in their waters the essential minerals required to maintain the fertility of the land. Briefly, in these river basins it now became possible to crop the same field year in and year out without having to allow it to lie fallow and without having to cut new fields into the forest. Indeed, within a few centuries mankind had in all probability completely transformed the appearance of these valleys.

In theory, with fields that could be cropped perennially it now became possible for the size of individual communities to grow. The village that until now had contained at most a few hundred souls could now contain a few thousand. But in the ability to expand there lay a danger. Were all the land available for agriculture in the immediate vicinity of a community given over to that purpose, the livestock would have to be fed in the woodland beyond, and between them cattle and sheep could, if contained within too small an area of woodland, reduce it to a waste in a very short period of time. Cattle are essentially branch feeders, eating the green leaves of trees and occasionally the bark, whereas sheep are ground feeders, nibbling at the newly growing shoots and undergrowth. Thus, cattle denude the standing forest and sheep prevent regeneration. Once forest had been destroyed by the grazing of herds and flocks there was an ever-present danger that the land would be given over to agriculture. Thus, two small villages which were initially at a comfortable distance apart might, if the population grew too numerous, ultimately find themselves in dispute, not so much over the land held for agriculture as for the residual forest used for grazing. Alternatively, all the land between separate villages might have been used for agriculture while the herdsman had to look farther and farther afield, to such an extent that the agriculturalist and herdsman became essentially distinct communities.

This state of affairs was not reached, of course, in a matter of a few years, and not necessarily in a matter of a few centuries, but it was ultimately to result in an extremely dense population in these river valleys.

Even from the outset, however, the same system of agriculture could not be followed exactly in both the Nile and Euphrates valleys, for the two rivers flood at different times of the year. The Nile floods in the early spring and the date at which it does so lies within very narrow limits year after year, a matter of a few days. Thus, the river can be allowed to flood, and once the waters have receded crops can be planted; it remains then only to ensure that the growing crops get enough water. The Tigris and Euphrates, on the other hand, are in spate in the early summer, and once their waters have receded it is too late to start planting. Sowing, therefore, has to take place before the rivers flood, with the result that there is a very real danger that the whole crop may be inundated and completely lost. If agriculture in the Mesopotamian valley, therefore, were to be a success, the fields had to be defended with dikes against flooding. Early written sources show us that flooding was an ever-present menace to the farmers of the Mesopotamian valley, while some authorities believe that the story of Noah and the Flood was a legend that the Hebrews acquired from the Mesopotamians. Unfortunately we know very little about early agriculture in the Mesopotamian valley, for later work has almost totally obscured whatever dikes and drainage canals were built by these early farmers. It is not until we come to written sources that we fully begin to appreciate the amount of irrigation work that had to be carried out by these people if agriculture were to succeed.

While in Mesopotamia dikes had to be built to prevent flooding, in Egypt canals had to dug in order to water the land after the floods had receded. Again it is not until we come to written sources telling of its first legendary king, who is reputed to have organized an enormous system of canals, that we get any firsthand evidence of the methods of irrigation at this early date in Egypt.

Grain crops were not the only plants grown by these people in the river valleys. The occasional find, from tombs and villages, of the remains of food show us that grapes, olives, figs and dates were all being cultivated. Mankind was thus becoming not only a farmer but a gardener as well. Thus, with fields that could be cropped repeatedly year after year, more long-standing policies such as the establishment of orchards which could be included in the general scheme of irrigation became feasible.

Before the year 4000 mankind had shown an interest in two metals other than copper: gold and silver. Unlike native copper, gold is not found in association with ores, but as veins usually in quartz rocks. When these gold-bearing rocks become weathered and eroded, the metal with the other mineral detritus is carried away by streams. The larger fragments of gold—the metal being, relatively, a very heavy mineral—are seldom carried far by water and so become deposited in the alluvial gravels of the upper reaches of the rivers. These so-called placer deposits often contain nuggets of gold the size of a pea or even larger, while farther down-stream only the finer particles will be found distributed as dust through the river sands. Theoretically, then, gold in the form of placer deposits had been available to man since his remote origins, and it may even seem surprising that it had not earlier attracted his attention. On the other hand, while the mountainous upper reaches of rivers are unlikely to have been particularly attractive to the hunters of big animals, the foothills surrounding the high-land plateaus on which early farming communities were so often situated may have been sporadically visited from time to time by shepherds. Vein gold—that is to say the metal embedded in quartz rocks—was clearly not being worked at this early period, for it required a great deal of heavy equipment to remove the metal from its matrix; one can only assume that placer deposits were being panned for the metal, a fairly simple process in which some of the sand or gravel was placed in an open dish with water and swirled around so that the lighter detritus slopped over the edge leaving the metal, if any, in the bottom of the pan.

Gold, unlike copper, does not harden when hammered and it is possible to beat the metal out into relatively thin sheeting without the need to reheat it from time to time to prevent its cracking. Furthermore, unlike copper, gold can be made to join only by hammering two pieces together, when a perfectly good weld will form; nevertheless, early objects of gold are invariably very small, and it seems probable that this is a reflection of the rarity of the metal rather than that this property had not yet been discovered.

Native silver occurs far less commonly than gold, although often the two metals are found together as a natural alloy, electrum, in which the proportions of the two metals vary rather considerably from area to area. Today these natural alloys would be classified as gold and the two metals would be separated, but to early man they were worked as they came, with the result that without chemical analyses it is often difficult to know whether small objects from the remote past were made of pure native silver or of merely an electrum containing a low quantity of gold (although a few pure silver objects of this early date are known and it is quite clear that early man was on the lookout for this metal too).

One other brightly colored rock was evidently much prized by these early farming people: the blue mineral lapis lazuli with its vivid ultramarine tone. Its natural occurrence is extremely rare and its source is still far from clear, but we know that it was in short supply and that at some time before the year 4000 attempts were made to produce a synthetic lapis lazuli. This synthetic material, usually referred to as Egyptian faïence because such enormous quantities of it were later produced in Egypt, can be looked upon as man's first real move into the world of synthesizing the material he required. In fact, the blue color of lapis lazuli is due to a highly complex molecular arrangement and the material was not truly synthesized until the nineteenth century. Early attempts at reproducing it were based upon the blue color that can be derived from copper when it is incorporated into a glass. The early examples of Egyptian faïence appear to have been made by powdering the surface of talc stone with one of the ores of copper—probably azurite or malachite—and then heating so that the whole surface of the object became a blue-colored glass. In

FIG. 48: *A tablet of faïence from Egypt; before 2000 B.C.*

Beads and small ornaments were made in
imitation of the blue stone, lapis lazuli, by heating
shaped pieces of soapstone in the presence of ores or
copper, which in effect covered the objects with a
turquoise glaze. Later the soapstone core was replaced
by a synthetic material (possibly mankind's first)
made by heating quartz sand and soda until the sand
particles fused. This material is generally known as
"Egyptian faïence." The work was probably carried out
in fires similar to those used for softening metallic
copper.

order to accomplish this a number of conditions had to be provided. First of all, the material could not be made in an open fire;
the work had to be carried out in a crucible or some other lidded
vessel which would keep the smoke and ash away from the object
being made. Secondly, a much higher temperature than is normally
found in the domestic fire was required. In short, the fire had to
be fanned or blown, and although we have no idea precisely how
this was done, at a very much later date we find metalworkers in
Egypt, for example, attaining these high temperatures by blowing
on the fire through hollow tubes; it seems highly likely that even
at this early date a reed was used to force air into the fire and so

raise the temperature sufficiently to allow the work to be carried out.

Egyptian faïence is still being made today in some of the less accessible parts of the Near East, as for example in Iran. The process is seen in its simplest form where the articles to be glazed are packed into a lidded clay vessel surrounded by a mixture of lime and wood ash to which has been added a small quantity of copper carbonate. The vessel is then heated for one day to a temperature of 950° C.—just below the melting point of copper— and on cooling the objects are removed, covered with a somewhat uneven blue glaze.

The importance in the history of technology of the appearance of Egyptian faïence does not lie in the immediate effect that the

FIG. 49: *Faïence-making in Persia today.*

"Egyptian faïence" is still being made in remote parts of Persia today. Although the process has obviously changed in details over the ages, it is probably basically the same as that used in 4000 B.C. The objects to be glazed are heated in a lidded clay vessel packed in a mixture of wood ashes, lime and green copper ore.

material had upon those living at that time—in fact, it probably had no worthwhile effect at all—but it did provide a set of conditions that may ultimately have led to the smelting of copper from its ores.

At about 4000 B.C.—in fact at much the same time that we find Egyptian faïence appearing in Mesopotamia, long before it was made in Egypt—we find also the appearance of true pottery kilns. As has already been pointed out, for polychrome pottery to be a success it was essential to separate the vessels from the fuel during fire. No doubt a number of different arrangements were tried, and indeed some of the early kilns show very considerable variation. But shortly after 4000 we find throughout Mesopotamia a quite standard form of kiln arrangement. The fire was lit in a hearth underneath the kiln chamber and the vessels were separated from the fire by a floor which was perforated in a number of places. Unfortunately, however, all the kilns so far excavated have been severely damaged and the upper parts are generally missing, but we would probably be safe in assuming that they resembled the kilns that we know of some 500 years later, which appear to have had a dome-shaped cover to the chamber with a vent at the top. These structures were built very largely of clay, usually with an outer wall of either stone or mud brick. As one might anticipate, a result of firing pottery in these kilns was that the structures themselves also became baked hard and are often far better preserved than the buildings around them made only of unfired brick. Apart from the obvious advantages of being able to separate fire and vessels, the kiln was also presumably less extravagant on fuel, and furthermore allowed a larger scale of pottery production.

Possibly to meet the growing demand, we now find a new tool introduced in the manufacture of pottery. Up to this time all pottery had been made by a number of simple means: by molding slabs of clay into preformed molds, by building up the walls of the vessel with successive rings of clay, or by some other rather tedious method which required a considerable amount of hand modeling. The new device that was introduced shortly before 3500 B.C. has sometimes erroneously been called a potter's wheel, but in fact it bore little relationship to what we know as the wheel today. More correctly it should be called a turntable. Its function was quite

(a)

(b)

(c)

FIG. 50: *The spread of "Egyptian faïence": (a) up to*

3000 B.C.; (b) 3000–2000 B.C.; (c) 2000–1500 B.C. The earliest objects of "Egyptian faïence" were made about 4500 B.C. in the Tigris-Euphrates valley. By 3000 B.C. they are found in the Lower Nile valley and in northern Iran, where they are probably imports. A thousand years later they are found more generally throughout the Near East and as far afield as the Indus valley. By about 1500 B.C. beads of "Egyptian faïence" occur all over Europe, where they were undoubtedly acquired by trade with the Near East. By this time Egypt has a thriving "faïence" industry of its own.

simple. Essentially it was a flat disk that could be made to rotate around a central pivot. A ball of clay was placed in the center and while the disk was made to rotate with one hand the other was used to shape the clay into the form of the required vessel. Unlike the modern potter's wheel, this central disk obviously could not be made to rotate continuously, but nevertheless it greatly speeded up the method of production and furthermore introduced a greater uniformity than before into the style of pottery being made within any single workshop. The turntable remained the chief means of pottery production for many hundreds of years after this time, and in fact we shall discover that it is not until perhaps 700 B.C. or even later that it was superseded by a true wheel that would rotate continuously.

The turntable and the kiln between them set new demands upon the potter insofar as his clay was concerned, for the body of his pottery could no longer be as rough and as coarse as before. Large, coarse particles incorporated into the body material would have tended to get caught in the potter's hands as the clay rotated on the turntable and, once dislodged, would cut through the walls of the pot he was making. The material had thus to be of far finer texture, and it had, too, to be more uniform, for clay that is in-adequately mixed can only be used as long as there are large, coarse particles in it to prevent excessive shrinkage and to make it porous during firing. But a clay containing fine particles that is poorly mixed will shrink unevenly as it dries, and furthermore lose water at different rates during firing, with the result that steam will build

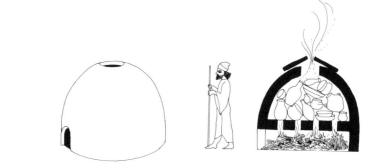

FIG. 51: *Reconstruction of a Mesopotamian kiln of about 3500 B.C., based on a number of excavated examples.*

The decorative effect of painted pottery would have been spoiled had the wares been fired in an open fire, for ashes and pieces of wood lodged on the surfaces would have caused smudges. The use of kilns, in which the vessels were placed on a perforated clay floor above the the fire, overcame this difficulty. To judge by the small number of kilns so far excavated, in Mesopotamia they were usually low, domed structures with a vent at the top.

FIG. 52: *Reconstruction of an Egyptian kiln of about 3000 B.C., based on a number of tomb paintings.*

By contrast, early Egyptian kilns, which we know of only from tomb paintings, appear to have been taller, chimneylike structures, open at the top. In some cases a platform was built to one side to allow the potter to stack his wares through the top opening. How the pots were set is a matter of conjecture: they may have been piled in haphazardly, or projecting platforms may have been built on which they could stand. The opening at the top appears to have been partially closed with a capping of mud or stone.

FIG. 53: *Egyptian potters at work: a tomb painting of about 2000 B.C.*

FIG. 54. *Reconstruction of a potter's turntable of about 3000 B.C., based on surviving fragments from Mesopotamia and on Egyptian tomb paintings.*

In the scene from an Egyptian tomb painting one of the tall kilns can be seen on the right being loaded with pottery. On the left is a second kiln being fired. Two men can be seen treading clay to make it homogeneous, while the potter can be seen squatting at a low turntable on which he is making his wares. Next to the turntable is a pile of prepared clay, and beyond stands an assistant ready to remove the vessels as they are made. The turntable was probably made of wood or clay and was pivoted into a socketed stone at the base. A mound of clay was piled on to the turntable, and from the top of the mound the potter shaped his wares, each vessel being cut off and handed to the assistant when finished.

up in the walls of the pottery which may cause the pot to burst, setting off a chain reaction in the kiln. To meet these new conditions the potters developed the practice of levigating their clay— that is to say, they mixed the raw clay thoroughly with water, allowed it to settle, then tapped off only the finer clay fraction from above, leaving unused the coarser lumps that had settled out at the bottom of the levigation pit. Thus, by 3500 B.C. the making

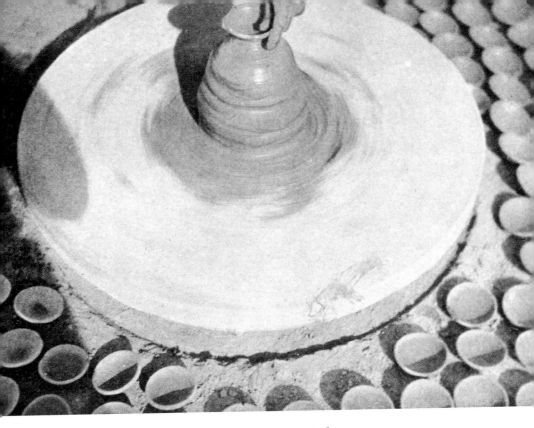

FIG. 55. *A modern potter, Sind.*

This system of shaping pottery can still be seen in some parts of the world today, for example in Sind, although here the wheel is larger and set lower than that in the picture from Egypt.

of pottery was beginning to become a highly complex industry requiring a considerable amount of know-how.

One small device appears at about 4000 B.C. which is of considerable interest in the history of technology and ultimately was to lead to one of the most important inventions of all time. This was the seal. Early seals were small circular disks of fired clay or stone into which was cut an impression of the mark required—usually some geometric pattern—while on the back of the disk was a small perforated lug used as a handle when making the impression. The device was clearly an owner's mark much like the seal in medieval and later Europe. It was, one imagines, usually impressed into wet clay, as, for example, when sealing jars or bottles, and, although ultimately the seal was to lead to a form of writ-

FIG. 56: *The back and face of a stamp seal from Mesopotamia;
about 3000 B.C.*

Small seals, usually of stone, were used to stamp the
owner's mark into wet clay as, for example, when
stoppering jars and bottles.

ing, for the moment it is important to notice that the idea of
producing a design in negative was fully appreciated by this time.

The stage was now set for two most important discoveries: the
fact that metallic copper could be reduced from its ores, and that
once reduced, molten copper could be poured into molds. A great
deal has been written about these two discoveries, and most of
what has been said has been speculation. On the whole it would
seem unlikely that we shall ever know precisely when and where
these steps were taken. Certainly, copper ores could easily have
been reduced to the metal during the process of the manufacture
of Egyptian faïence or even during the manufacture of pottery,
although the former seems more likely. The process requires of the
metalworker some understanding but not a great deal of equipment. The ore was mixed with the fuel—ideally charcoal, although
any very dry wood would have sufficed—and placed in a shallow
pit in which a small fire had already been kindled. The fire then
had to be brought to a white heat, using the blow tubes already
mentioned. The high temperature had to be maintained for a considerable period, probably the best part of a day, and then the
whole fire was allowed to die down. When cool, the molten copper
would settle to the bottom of the pit, while above it lay a layer of
glassy slag that would be chipped away and discarded. The metal
as it stood would be full of blow holes and very unpleasant to work,
although with a great deal of hammering and annealing it could
have been shaped into objects of various forms. On the other

FIG. 57: *Reconstruction of a simple mold for casting spear- and axheads.*

FIG. 58: *A copper axhead that would have been cast in such a mold, from Palestine; about 3000 B.C.*

With the discovery that copper would become fluid at sufficiently high temperatures it became a practicable possibility to shape the metal by pouring it into molds. The concept of casting already existed as in the making of mud bricks (Figure 26) and in the seals above. At first molds used for casting copper objects were simple depressions cut into the surface of a suitable stone. Objects cast in such molds still needed a great deal of hammering and polishing to bring them to the required shape.

hand, if the ingot of blistered copper were broken up, placed in a crucible and reheated in a small furnace of the type just described, it would become fluid and could be poured out when the worst of the blistering disappeared. At this time the mold, admittedly used for the making of bricks, was very nearly two thousand years old, and stamp seals, as we have just seen, were already in use. It is hardly surprising, therefore, that at the same time that we find early man reducing metallic copper from its ores we also find him

FIG. 59: *Reconstruction of a two-piece mold used for casting more complex shapes.*

FIG. 60: *A tanged knife that would have been cast in such a mold, from Palestine; about 3000 B.C.*

By using two opposing mold pieces it was found that a great deal of the final shaping by hammering could be avoided. The molds were made either of stone or of fired clay, the two pieces being held in the correct position by means of short dowels.

designing molds specifically for the casting of that metal. At first these were very simple devices. The shape of the object to be cast was cut as a negative into a suitable piece of stone such as talc stone or a fine-grained sandstone—indeed any rock that would stand the high temperatures involved—and into this depression was poured the molten metal. Thus, initially, no attempt was made

FIG. 61: *Diagram showing the use of a false-core ax mold.*

FIG. 62: *An ax from Mesopotamia that was cast in such a mold.*

A further development of the piece mold was the incorporation of an additional clay core. By this means it was possible to cast socketed bronze objects such as the ax shown in this diagram.

FIG. 63: *Egyptian metalworkers from a tomb relief of about 2500 B.C.*

FIG. 64: *Diagram showing how the crucible was probably held.*

> With the discovery that copper could be extracted
> from its ores, the metal became more plentiful. This
> scene from an Egyptian tomb shows copper being
> melted on the left, and on the right being poured from
> a crucible into the mold. In this case it almost appears
> that the red-hot crucible is being held directly in the
> hands of the founder. Probably it was grasped between
> two stones as shown in the accompanying drawing.

to produce in detail the shape of the object required. Instead a
blank was formed which later would be hammered and annealed
until the required shape was achieved. But within a span of a very
few centuries new methods of mold designing were being evolved.
The smith had learned to cut molds of two opposing pieces bearing
the larger part of the required design of the finished object so that
when the two pieces were held firmly together the metal would be
poured into the space between them in such a way that when cool
an almost perfect form of the object had been produced. Furthermore, the two-piece molds were often no longer made of stone but
of fire clay. In such cases the smith began by making a pattern of
the object he needed to cast, often of wood, and around this pattern
was built first one, then the other, piece of the clay mold. The
pattern was then removed, the mold pieces were fired like pottery,
reassembled, and into the space once occupied by the pattern was
poured the molten metal.

While the earlier form of mold—the simple depression cut in
a piece of stone—allowed only limited shapes to be produced,

FIG. 65: *A helmet of sheet gold and a cow of copper: fine examples of Mesopotamian metalworking of about 2500 B.C.*

> Despite improved casting methods, the hammer still remained the major tool for shaping metals. This helmet, for example, was made entirely of hammering, whereas the decoration was added by chasing the surface with punches.

since the upper surface in the mold always had to be flat, the later piece molds allowed a far more elaborate series of objects to be manufactured. With increasing supplies of copper from the process of smelting, and with improved methods of casting copper, objects made of it now became quite common, although the metal was probably still only available to very wealthy people, and hence apart from ornaments and trinkets we find the use of copper limited almost entirely to the manufacture of weapons. It was still too rare to be used extensively for the manufacture of tools.

With the extensive use of kilns for firing pottery and furnaces for making metal objects, it is hardly surprising that by 3500 B.C., in Mesopotamia at least, experiments were made with the firing of bricks. Hitherto all the building in this area and in Egypt had been carried out in sun-dried mud brick, and although this allowed the construction of fairly large buildings it had the very great disadvantage that in the winter frost and rain caused the surface to disintegrate. Losses on the surface of walls due to weathering could of course be plastered over with more mud from time to time, but in due course the building was bound to become both unsightly and unserviceable. In fact, at this point in time villages, particularly in the lower valley of Mesopotamia, can no longer be looked upon as simple settlements serving as markets but must be seen as embryonic cities with an administrative body that insisted on being housed in a suitable style. Thus every small city had its own central temple, and it was apparently the temple servants who became the administrators. A dilapidated mud-brick temple in the middle of the city would, of course, have been a contradiction in terms, and it was to these often quite substantial buildings that the new technique of using fired brick was applied. Early forms of sun-dried mud brick had normally been on the rather large side,

FIG. 66: *A brick vault from Ur, Mesopotamia; about 2500* B.C.

Fired bricks, of smaller dimensions than the sun-dried
mud brick, were used to a limited degree especially to
provide decoration or to give added strength to the
building. The use of fired brick made possible the
development of true arches as seen in the vault of this
tomb from Ur. For reasons of economy however,
sun-dried brick continued to be used for the major part
of all buildings.

as a rule measuring about eight by sixteen by two and a half inches.
Such large bricks clearly proved too difficult to fire, and we now
find that a far smaller brick was used, very similar in proportion to
our own modern bricks. Furthermore, various shapes of brick were
also made, as, for example, slightly curved forms for building pillars. But it would be quite wrong to imagine that the whole temple
was made of fired bricks, for often they were used only where one
would expect a great deal of wear and tear. A second, quite different solution was found in order to prevent the excessive weathering
of sun-dried mud-brick walls. Into the relatively soft surface of

FIG. 67:
Cone mosaic from
Mesopotamia; about 2500 B.C.

FIG. 68: *Diagram showing how the cones were hammered into the*
soft mud-brick wall.

Cones of fired clay or of stone were hammered into
mud-brick walls to produce decorative effects. Walls
treated in this way were naturally far less prone to wear
and tear than those of unfired brick.

the wall were hammered a number of cones, some of which were
made of fired clay, others of differently colored stones. These
cones were usually set into the wall to form a pattern, their exposed
heads making a façade that gave the appearance of a mosaic.

The administrators of the city's affairs were, however, faced
with a far greater problem than that of simply keeping the temple
in good order, for each city now controlled a sizable acreage of
land and many cities had quite a number of dependent villages.
To administer such a community properly it was found necessary
to be able to keep records of such things as the taxes to be paid and
the services to be provided by individual citizens or smaller dependent communities. We have already seen that the stamp seal

pressed into clay was being used as a means of identification by 4000 B.C. and then five hundred years later we find the palace servants adapting this idea as a means of keeping records. In brief, we find the beginnings of a system of writing in which signs were made on slabs of wet clay which were dried and stored as a record. The symbols used were not with a stamp, however, but were drawn freehand. Initially the symbols used were pictograms. That is to say, if one wished to record the payment of a sheep then a picture of a sheep was drawn. But if confusion, of the sort so adequately described by Kipling in his story of the First Letter, were to be avoided, then agreement had to be reached on how a sheep was to be drawn so that it could not be confused with any other animal. Indeed, in the earliest period of recording, we find several different symbols used to denote sheep of different kinds—lambs, yearlings, ewes, rams, and so on. Later a single symbol was used to denote all sheep, and a series of other signs was appended to indicate what type of sheep was meant. In the course of time these pictograms became greatly abbreviated and simplified to such an extent that without a record of the various stages of abbreviation it is often difficult to see any resemblance between the symbol and the original pictogram.

FIG. 69: *The earliest form of writing on clay tablets from Mesopotamia; about 3000 B.C.*

FIG. 70: *A cuneiform tablet from Mesopotamia of about 800* B.C.

With the growth of cities in Mesopotamia the administrators found it necessary to keep records. At first they were made by sketching commonly understood symbols (ideograms) on the surface of tablets of damp clay. These, when dry, were stored in cellars, so that many have survived.

In time the practice of sketching the ideograms gave place to a method in which each figure was built up by impressing a reed of triangular section into the surface of the clay to give, ultimately, the form of writing known as cuneiform. The ideograms bear no immediate resemblance to the sketches from which they were derived.

Drawing easily decipherable symbols with a pointed object on the surface of a piece of wet clay is not an easy matter, and in Mesopotamia the ultimate solution was found not in the making of freehand drawings, but by impressing into the surface a small stylus cut from a length of reed which was itself triangular in section, and which thus made small wedge-shaped impressions in the clay. The freehand symbols were further modified to this system of writing so that every symbol was now composed of a series of differently positioned wedge-shaped impressions. The end product seems even more remote from the original pictogram than ever before. In the subsequent centuries this system of writing, referred to as cuneiform, spread throughout a larger part of the Near East, although it was never greatly used in Egypt, where a different writing material—as we shall see later—led to a rather different system of recording.

From our point of view the records from the temples of the various cities of Mesopotamia come as a salutary reminder of what we have overlooked in this account of early technology. On the clay tablets are depicted a number of objects, and even pieces of equipment, of which there is little if any record in the way of material remains. Two such pieces of equipment are the plow and the cart, both, of course, drawn by beasts of burden. Unfortunately

FIG. 71: *Cattle yoked by the horns, from an Egyptian tomb painting; about 2000 B.C.*

Oxen were probably the first animals used for pulling sledges and plows, the yoke being lashed to the animals' horns rather than placed over its shoulders. How the yoke was tied to the horns of a pair of oxen can be seen clearly in this painting from an Egyptian tomb which, although late, probably depicts the earliest form of yoke.

FIG. 72: *Two types of shoulder yoke for oxen from antiquity.*

The yoke placed over the animals' shoulders was, however, far more efficient since it allowed the oxen to exert a greater pull. The ox yoke, once placed over the animals' shoulders, required no further harness to keep it in place. Of the two types of shoulder yoke, that on the right is clearly an element of the pictogram of plows in Mesopotamia (Figure 120). The other type appears to have been used commonly in Egypt.

we remain far from sure precisely what draught animals were first used or, indeed, how they were harnessed. Horses, which are naturally steppe dwellers, were certainly not domesticated in Mesopotamia at this period and it seems likely, therefore, that the earliest beasts of burden were the ox and the wild ass, or onager; of these two it would appear probable that the ox was the first to be made to draw man's vehicles. This is suggested only because the earliest illustrations of plowing in Egypt show oxen harnessed to a plow not by a yoke but by a rope tied directly around the animal's horns; and the yoke—and, by inference, the use of onagers—for pulling vehicles would seem, therefore, to be the later of the two. Inefficient, and even cruel, as the harnessing of oxen by their horns may seem to us today it could well have been used as a means of shifting heavy weights long before the introduction of either the plow or the wheeled vehicle, and the earliest pictogram inscriptions from Mesopotamia show yet a third vehicle: the sledge. Indeed, the pictogram used to describe the wagon is simply a sledge with two pairs of wheels attached. Sledges of various types continued to be used in the Near East for moving very heavy weights, for a very long time indeed, as we shall see later.

FIG. 73: *The remains of solid wooden cartwheels found during excavations at Ur; about 2500 B.C.*

It is still far from certain where the first wheeled vehicles were developed. The first examples known come from Mesopotamia. Here the wheels are made of solid wood, but they are not made of one single piece. Instead, they were built up from three sections, so it would appear that there may well have been earlier wheels, made of a single piece of wood, about which as yet nothing is known.

In the five hundred years that follow 3500 B.C. we know of a limited number of two-wheeled carts and four-wheeled wagons as actual vehicles, as pictures, and as small pottery models. The wheels of these vehicles were made of solid wood, but even in the earliest forms in which we know them they were never made of a single piece of timber but always of three planks joined by two cross struts. The middle plank, which was initially larger than the other two, had been shaved away to form a heavy central hub and to this on each side were attached by the cross struts two partly circular planks. The fact that this pattern of wheel was copied in other parts of the world to which it spread—although the construction of the

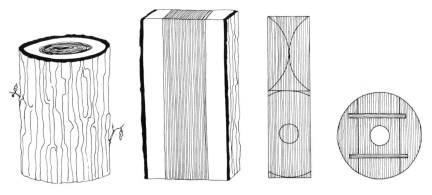

FIG. 74: *Diagram showing the process of making a three-piece wheel.*

FIG. 75: *An Irish cart, County Tipperary.*

Wheels made from three sections of planking are still made in many parts of the world today. Usually a plank is cut from the center of a log, and the outer, soft sapwood, which would wear rapidly, is removed. The plank is then halved: one half forms the middle plank of the wheel; the other half is further divided to form the remaining two sections.

FIG. 76: *Limestone relief from Mesopotamia of about 3000 B.C., showing a cart drawn by onagers. The wheels were made in three sections.*

In Mesopotamia wild asses (onagers) were used to pull carts as well as oxen. The yoke did not fit comfortably over the onagers' shoulders and had to be held down by a neck strap, which probably had a tendency to choke the animals whenever the yoke rode up. To guide the onagers copper rings were put through their noses, to which reins were attached. Although in Mesopotamia carts with two and four wheels were in use, no such vehicles appear to have been employed in Egypt where, one supposes, boats and rafts on the Nile were the main means of transport.

carts and wagons themselves differed widely—suggests that the idea of wheeled vehicles diffused from a central point—namely, Mesopotamia. It also seems that the axles were not attached permanently to the vehicles but were held in place by straps or some other device so that they could be removed readily, a feature which suggests that it was not an uncommon event to have to take one's cart to pieces when the going got too rough, and to reassemble it once the obstacle had been overcome. It follows that early carts and wagons were not used for long-range transport but rather for shifting heavy loads over short distances, either from outlying villages to the central city or perhaps even from one adjacent city to another.

Even so, it must be admitted that there may be a very considerable period in the early development of the wheel about which we know nothing. On the face of it, we would expect the earliest wheels to be of a single piece of wood similar to those so often depicted by humorists, in which case the wheel may well have been developed outside Mesopotamia, even perhaps as far away as the steppe lands of Asia. That this was so is to some extent suggested by the fact that solid wheels made of a single piece of wood do turn up in Western Europe at a very much later date. Indeed it seems likely that in Europe the wheel was borrowed from some center other than Mesopotamia, as, for example, southern Russia.

With the introduction of the yoke rather than attachment by the horns it became possible to use the onager as a draught animal, but as a means of guiding it the bridle and bit were still unknown, and the reins were attached to a ring of copper passed through the animal's nose—similar to the ring still used today for controlling bulls. The yoke was conceived as a cross member to a single draught pole, and this meant that for any vehicle there had to be a pair of animals. Furthermore, the onager's shoulders, unlike those of the ox, are not ideally suited for a yoke and it became necessary to pass a collar around the animal's throat to keep the yoke in place, as well as a girth band to keep the yoke down on the animal's shoulders. The effect of the breast collar was partially to choke the animal, particularly if it exerted itself greatly, and this tended to reduce its efficiency. The onager, at any rate, was a fairly small animal and therefore it is not uncommon to find that not two but four onagers were used, the outside pair having been attached by their collars only to the two ends of the yoke. The introduction of paired draught was in the long run to have a very considerable effect upon the history of the development of wheeled transport, for it seems that, once having introduced the single draught pole and the yoke, the intellectual effort required to change this means of traction was too great, and indeed within the span of this particular history we shall barely see the introduction of either the shaft or the horse collar. Thus, in the ancient world there was no such thing as efficient heavy horse transport.

It is quite probable that the introduction of pair draught was the outcome of using beasts of burden to pull plows, for the plow existed long before any form of wheeled vehicle, but as it was con-

FIG. 77: *Pictograms of plows from Egypt of about 3000 B.C., with a reconstruction based on them and on later pictures of plows.*

In both Egypt and Mesopotamia early forms of plows were little more than a forked branch dragged through the soil by a pair of oxen, the plowman holding the two branches of the fork as handles, while the junction, sharpened to a point, served as a primitive share. The plows have long since disappeared, but their shape can be deduced from the pictograms used to record them in both areas, as well as from later pictures (Figure 119).

structed entirely of wood we have no early plow remaining from excavated sites. It is nevertheless interesting to notice that while in many of the older civilizations the invention of the plow was attributed to a god, the introduction of the wheeled vehicle does not appear to have received the same divine attribution. Anyway, the earliest plows were probably pulled by a team of men and women, and indeed in some of the earlier Egyptian tomb paintings we have pictures of just such operations.

The earliest form of plow that we know of amounts to little more than a forked stick, the two tines of the fork serving as handles and the junction as the share. Initially a rope was attached just above the junction and on this rope the plow team pulled while the plowman pushed down on the handles. The effect was to produce a rather narrow furrow in the soil. Before plowing, the soil had to be broken up, and although most of the soils cultivated at this time were fairly light, we find that not only was the hoe used for this purpose but also the larger clods of earth were broken up by mallets. Into the furrow made by this rather primitive device the sower put in his drill of seed. In time the towrope was replaced by a draught pole secured to a yoke at one end and lashed at the correct angle to the V-fork at the other. Indeed, this type of plow survived with remarkably few modifications almost entirely throughout the history of ancient Egypt, although the lashings between the draught pole and the share were ultimately replaced by wooden cross braces while the handles were often braced with wooden struts. Although early Mesopotamian plows appear to have been of this kind, by about 3000 B.C. we find the introduction of the first major modification. A single pointed piece of timber formed a share and sole, the share actually cutting the soil while the sole helped to push it aside, so creating a deeper and wider furrow.

By our standards these early plows must be seen as hideously inefficient, for in practice they failed to do more than just scratch the surface of the soil; nevertheless, they must have brought about vast improvements in crop yield, for not only did they allow sowing to become consistently even, but since the crops were in regular rows they allowed efficient weeding. Of all the devices created by mankind up to the end of the fourth millennium B.C. there can be little doubt that the plow had the greatest over-all effect. More probably than not it was mainly responsible for the rise in population in the small Mesopotamian and Egyptian cities.

4/Of Monuments, Ships, Metallurgy and Military Technology

(3000–2000 B.C.)

THE TWO OR THREE CENTURIES centered around the year 3000 B.C. appear to have been fairly critical in the history of the development of early technology. This view may be fortuitous, for not only from this period do we begin to get Mesopotamian written records that we can comprehend, but also the political and religious situation in both Mesopotamia and Egypt gave rise to elaborate forms of burial, and from the many tombs known to us we can derive a vast amount of information. The rulers in these countries as well as the more important citizens elected to be buried in capacious graves with all the goods and chattels that they imagined they might require in the next world, while at a slightly later date there began the series of complex tomb paintings in Egypt that were to furnish us with so much valuable information. So, what seems to us to be an era of sudden technological advance may appear so only because we have a dramatic increase in sources of information.

In the lower Mesopotamian valley there had evolved two small kingdoms. In the immediate area of the delta were the Sumerians, of whose cities Ur is probably the best known, while farther to the north were the Akkadians, occupying an area in the Euphrates valley with Babylon as its approximate centre. In both Sumer and Akkadia there were several cities, each with its own ruler, and although they were often at war with each other, in neither kingdom was political unity achieved. Nevertheless both regions evidently became extremely wealthy, and this wealth must have depended entirely upon the products of agriculture and husbandry, for apart from their crops and livestock and an almost inexhaustible supply of mud, the countries were quite devoid of natural resources, all other materials of necessity having to be imported. Even so, from the graves of the rulers of these small cities have come many objects which show that considerable technical advances were being made.

In the valley of the Nile a somewhat similar development appears to have taken place, but even before 3000 B.C. we find that the cities had formed a loose political alliance and that the country as far upriver as Aswan was divided into two major units—the one, Lower Egypt, that area of the delta and the Nile valley as far south as Memphis; and the other, Upper Egypt, between Memphis and Aswan. The two countries were ultimately to be unified by the ruler Menes, the first pharaoh, who established his capital at Memphis. Tradition has it that Menes not only concerned himself with the unification of Egypt but also with the control of the river: to him is attributed the first damming of the Nile, the digging of dikes for agricultural purposes and indeed the first attempt to control and apportion the waters of the river. The wealth of Egypt was thus, with Mesopotamia, based upon its agricultural output. However, unlike Mesopotamia, the Egyptians had on their doorstep a number of mineral resources that they were able to exploit with little effort, including copper ores, gold and a wide range of rocks suitable for building and the making of a wide variety of ornaments.

Somewhere shortly before the year 3000 metallurgists made a discovery that was to transform the entire "industry." They found that by mixing a small quantity of tin ore with the copper ores when they smelted them, a harder and altogether more useful metal than copper was obtained. Briefly, they discovered the alloy bronze. The occurrence of tinstone can never have been very widespread in the Near East and it is interesting to notice that, geologically speaking, it does not occur in the same type of deposit as do the ores of copper, but, rather, occurs in those areas in which we might expect to find veins of gold. Thus tinstone, which is a fairly dense mineral, may well have first been noticed during washing for gold; and metallurgists, finding that the little black lumps of ore were relatively heavy, presumably made various attempts at smelting them until they arrived empirically at a suitable alloy with which to make tools and weapons. Bronze, however, was not only a harder material than copper but it was also more easily worked, for the effect of alloying a small quantity of tin with copper is to reduce the melting point of the resulting metal, and in practice this meant that at the same temperature at which they were accustomed to cast copper the bronzesmiths found now that they had a far more fluid metal which was much easier to cast. With the introduction of bronze, therefore, the quality of casting improved dramatically.

It is most unlikely that this discovery was first made in Mesopotamia. In all probability it was made much nearer the sources of the metals, in the mountainous areas of Syria or eastern Turkey, for example, but the Mesopotamians had the wealth to purchase this new metal and the wealth to employ craftsmen to fashion it. Thus, it is from the tombs of the early Sumerian kings that we find the first examples of bronze used in any quantity. The situation in Egypt was totally different. Endowed with large quantities of copper ore, the country was utterly devoid of tinstone, and thus at this early period the Egyptians continued to use copper where in Mesopotamia bronze would have been employed; it was not for another thousand years that bronze became the common metal in Egypt. But despite the difficulty of working the pure metal, the Egyptian smiths achieved remarkable results with copper alone.

A study of the metal objects from the royal Sumerian tombs shows that the smiths had made very considerable technical advances. Complex objects were often cast in molds made of two, three, or even four pieces, and it is quite clear from the chemical composition of some of the weapons and ornaments that the smiths had been experimenting quite extensively with various alloys to such lengths that they had hit upon the idea of joining one piece of metal to another with an alloy of a different composition and had thus in effect begun to experiment with solders. But excellent though their technique of casting was, much of the work was still accomplished by laborious hammering and chasing, while the size of the castings they were able to make was still severely restricted and seems to have been limited by the quantity of metal that they could melt at any one time. Nevertheless, some fairly

FIG. 78: *Boat depicted on a jar from Egypt; before 3000* B.C.

FIG. 79: *Men building a reed boat, from an Egyptian tomb painting;*
about 2500 B.C.

FIG. 80: *Boat depicted on a stone vase from Mesopotamia; about*
3000 B.C.

FIG. 81: *Reconstruction comparing Egyptian and Mesopotamian boats of about 3000 B.C.*

FIG. 82: *Model of a boat in silver, Mesopotamia; about 2500 B.C.*

FIG. 83: *Boat used by Arabs in the Euphrates delta today.*

Well before 3000 B.C. pictures of boats appear on Egyptian pottery and on Mesopotamian seals, but without a knowledge of what was to happen in the following few centuries it would be impossible to guess what they were like or of what they were made. The Egyptian jar shows a boat with an exaggerated number of paddles, and perhaps a square sail.

The much later silver model of boat from Ur, however, gives a far clearer picture of this type of craft, while the Egyptian relief of much the same date leaves no doubt as to how and of what such craft were constructed. Reeds were tied into bundles, and then were further lashed, bundle upon bundle, to produce a boat with no keel and an upswept stern and bow.

The major structural difference between boats in Egypt and Mesopotamia appears to have been in the treatment of the bow and stern. In the former region the terminals were left blunted, the cut end of a single bundle of reeds; in the latter area the bow and stern were brought to a point. In all probability the Mesopotamian craft were covered with bitumen as are the boats of the marsh Arabs today.

large pieces of ornamental metalwork were produced by the extremely simple expedient of heating sheet copper over a wooden substructure to which the sheeting was attached by nails.

We saw in the last chapter how early carts and wagons were unlikely to have played a large part in the long-distance transport of imports into Mesopotamia, and it is now for the first time that we get any direct indication of the type of boat being used for this purpose in Sumer and in Egypt. In both areas the boat appears to have been made largely of bundles of reeds laid horizontally and lashed together. From Mesopotamian seals and their impressions, as well as a model boat in silver from Ur, it is evident that here the early vessel was built of bundles of reeds terminating with upturned, pointed bow and stern. Indeed, the silver model just referred to could just as well be one of boats used today by the marsh Arabs in the delta of the Euphrates, for they are built in exactly this manner. From Egypt we have one or two very bad pictures of early boats drawn on pottery from the predynastic period (before 3000 B.C.,) but these drawings become intelligible only when we compare them with the really superb tomb paintings of the later periods. The Egyptian boats were clearly built of bundles of papyrus, and although both bow and stern were raised, the bundles of reeds were cut off to give a blunted, upturned end. In both regions the boats were of course without keel and of shallow draught. They were also light, and this may have been an important factor in negotiating the rivers, for at the cataracts portage would have been essential.

In both Mesopotamia and Egypt the boats appear to have been paddled at the beginning of the third millennium; it was only later that the Egyptians took to rowing their vessels. The very poor drawings from the predynastic period in Egypt already mentioned would suggest that some of the early boats were being sailed, but at this stage of development a single mast evidently proved too difficult to step, and we find in use a bipod mast supported by an after stay rope. The sails were presumably of linen, each vessel having a single square-rig sail which was furled by lowering the yardarm.

Vessels constructed along these lines obviously had very severe limitations: if they were built too long they would clearly break their backs in rough water or when heavily laden; they had little freeboard and would thus easily capsize in a rough sea. They were, thus, essentially rivergoing craft, and it is not for another five hundred years that we shall find anything approaching a seagoing ship. Nevertheless, for the carrying of valuable materials in both Mesopotamia and Egypt there can be little doubt that the boat was the most important single means of transport. Heavier loads of course could have been shifted by raft, and although we may surmise that rafts were already in use—indeed reed boats were most probably developed from rafts—it is not until we come to the period of stone monumental buildings that we find reliefs showing the carrying of huge blocks of stone by raft.

The very different nature of the available raw materials in

FIG. 84: *An Egyptian vase showing a boat with sail; about 3000 B.C.*

By 3000 B.C. reed boats of the kind just described
were being sailed by the Egyptians. A simple square
sail, probably of linen, was set well forward. Very early
pictures show a single mast, although later this was
replaced by a bipod mast.

Mesopotamia and Egypt led to strongly contrasting methods of
building in the two areas. In Mesopotamia, as we have already
seen, brick, at least for the more important buildings, was already
being fired, and we now find in common use a type of brick apparently evolved to overcome a particular difficulty. A simple flat
mud brick tends to warp on drying and to bend still further on
firing. The Mesopotamian brickmakers developed the trick of
building up a dome of mud in the top of the mold, which prevented excessive warping but which gave a brick that was flat on

FIG. 85: *Building a wooden boat, as depicted on the walls of an Egyptian tomb; about 2500 B.C.*

By 2500 B.C. the Egyptians were certainly building boats of wood, possibly for many centuries before. But their method of construction was clearly inspired by the reed boat: planks were doweled or sewn, edge to edge, without keel or ribs.

the underface and convex on the upper surface. Using these bricks, the builders experimented with a wide variety of bondings including a herringbone arrangement. It was presumably experimentation of this sort that led to the development of the true arch, which first appeared at this period and enabled builders to span wide spaces without the need to import large stone lintels. The Sumerians, of course, did import a limited amount of fine stone for facing and for querns, but their buildings remained, nevertheless, essentially brick ones.

By contrast, the Egyptians were content to continue building with unfired mud brick for all domestic purposes, including royal palaces; because of this we know less about their normal domestic lives than we would otherwise know, for these buildings have long since been totally destroyed. For monumental building, however, they began to quarry and to dress sizable blocks of stone. Thus, the predynastic kings and those of the very early dynasties were buried in mud-brick houses of the dead surrounded by all the objects that they felt they might need in their afterlives. These mud-brick structures were too easily robbed, and to overcome this desecration the burial chamber was then sunk beneath the house, which itself became a completely stylized copy made of mud brick and rubble, although the burial chamber and the outer facing of the structure were sometimes faced with stone. Ultimately this arrangement of sunken burial chamber surmounted by a stylized house of the dead was to lead to the design of the pyramid as a suitable mausoleum for the pharaoh and his family.

The stone most favored for building pyramids was a fine limestone obtained from one of a limited number of quarries in the vicinity of the Nile itself. By this time the Egyptians had become fairly adept at working stone. Even in predynastic times they had learned how to make stone vases and other vessels out of harder rocks than the limestone of which the pyramids were constructed, so that as a building material it presented few technological problems to the Egyptians. Each block of stone appears to have been cut clear of the parent rock at either side and at the back by making a narrow draught, using either copper chisels and mallets or hand-held stone picks made of diorite. Along the line of what was to be the lower edge of the block a series of pits was then cut and into these were driven wooden wedges causing the block to split away from the parent rock. It should perhaps be pointed out that even today, although mechanical tools are used for much of the cutting, the same method of quarrying is still used in many parts of the world.

For the Egyptians a far greater problem was presumably presented in the shifting of these blocks of stone once quarried, for many of them weighed several tons. To transport the building stone down the Nile by barge was a relatively simple matter; the main difficulty must have been to carry the stone from the quarry to the river and from the river to the building site. Wheeled vehicles never became popular in Egypt until very much later, although the Egyptians of this period occasionally used wheeled assault ladders as a military device, as we shall see later. For transporting heavy loads overland the Egyptians seem to have preferred sledges, and there are several tomb paintings showing how heavy burdens were moved. Occasionally the sledges were pulled over rollers, but this was not invariably the case, and more often one feels that the future trackway for the sledge was carefully prepared beforehand. Water poured in front of the runners was used as a lubricant and a large team of men hauled the sledge by massive towropes usually made of papyrus.

Once the sledge had arrived at the building site there remained the problem of lifting the block of stone into position, but on this subject the tomb paintings give us no help. However, from various excavations we have sufficient evidence to suggest that a ramp was built against the side of the pyramid up which the sledge and its burden were drawn. The block of stone was ultimately levered into position from the sledge, and as the pyramid grew so did the ramp, which was removed after the last block had

FIG. 86: *Quarrying wedge slots, Aswan.*

> Stone for building pyramids was quarried by the same method as that still in use today, although the tools were all of stone, wood and copper. A line of slots was cut into the quarry face with stone picks or copper chisels. Wooden wedges were then driven into the slots, causing the rock to split along the line.

been laid. Although before being lifted each block of stone was trimmed carefully to fit precisely to its neighbors, the ultimate face-trimming was not carried out until after the building had been completed. For this work it seems highly unlikely that the mallet and chisel were used, but rather that it was done with hand-held picks of diorite or other tough stone, while for a really fine finish the surface would have had to be rubbed down with blocks of sandstone.

Although from the year 3000 B.C. onward bronze and, in Egypt, copper were used increasingly for the making of tools, one has to confess, nevertheless, that a large part of the production of these metals went into the manufacture of arms and weapons. The reliefs and wall paintings of this period obviously show an idealized situation, but even allowing for that it is perfectly clear that the rulers of the various states had well-equipped and well-trained armies. The backbone of the army was the infantryman, who was usually armed with a spear, a dagger and a large shield. Also many of the infantrymen wore helmets, but it is unlikely that these, or the shields of the rank and file, were made of metal. They were more probably leather. The Sumerians and the Akkadians, but not the Egyptians, also employed a number of wheeled vehicles, though it is far from clear what their function was in battle. Wagons with four solid wheels drawn by four yoked onagers seem to be mobile armories rather than chariots, and were probably not used in the fore of the battle at all. But there are other lighter two-wheeled vehicles, equally drawn by four yoked onagers, that may possibly have been used as chariots. Even so, their design suggests that they were fairly slow, lacked maneuverability and were exceedingly vulnerable. One can easily imagine that well-trained infantry would soon have learned to cope with them.

Apart from ensuring that one had an army as well trained and equipped as one's enemy's, the principal method of defense was to girt one's city with an adequate wall. In fact, city walls appear far earlier than this period. Even the pre-agricultural community at Jericho had set a defensive wall around their village, and many of the early agricultural villages were designed in such a way that their defense from sudden attack could have been an easy matter. Now, however, the tall city wall became *de rigueur*, while its lim-

FIG. 87: *A relief showing a phalanx of troops from Mesopotamia; about 2500 B.C.*

FIG. 88: *A tomb model showing a phalanx of troops from Egypt; about 2000 B.C.*

> The power of both the Egyptian and Mesopotamian rulers depended to a large degree on well-equipped and disciplined armies. For the rank and file, shields and helmets were made of leather. The main weapons were the spear and the arrow, both of which were most commonly tipped with stone points. Since only the most wealthy could afford metal arms, copper and bronze were reserved largely for officers and crack troops. The armies were composed almost entirely of infantry: spearmen and archers.

ited number of gateways allowed not only a manageable defense but also a means of regulating the comings and goings into the city. With one's enemy safely protected by the city wall the attacking army had little choice but to attempt an assault or to settle down and besiege the city. At this early period the only assault vehicle appears to have been a scaling ladder on wheels—known from an Egyptian tomb painting—and methods of breaching the wall do not appear to have been devised. Unless the city garrison were very small, one can only surmise therefore that the more normal method under these circumstances was to besiege the city.

In the two or three centuries immediately following 3000 B.C. the technological innovations that had taken place in Mesopotamia and Egypt were being slowly diffused into other areas of the Near East. Copper was being widely used in Anatolia and also in Cyprus and in Crete, and while bronze appears just as early in Anatolia as it does in Mesopotamia it is not until shortly before 2000 B.C. that it becomes commonly used in this area. The introduction of the knowledge of copperworking into Cyprus, however, was to have very considerable repercussions, for the copper ore deposits in that island were by any standards fabulous and a busy trade in the metal between the island and Anatolia and Syria was rapidly built up. At this point in time, too, the city of Knossos in Crete appears to have been established as a trading center, reflecting architecturally many of the aspects of the cities of the Near

FIG. 89: *Copper model of a chariot from Mesopotamia; about 2500 B.C.*

FIG. 90: *Reconstruction of this type of chariot based on the copper and contemporary clay model.*

The Mesopotamians were experimenting with chariots, however. These were drawn by four onagers, the inner pair yoked, the outer pair pulling on traces. The vehicles seem unduly cumbersome and, one feels, enemy infantrymen could have easily dealt with them. Probably they were used in battle to give the commanders greater mobility.

FIG. 91: *An Egyptian wheeled scaling ladder, from a tomb painting; about 2500 B.C.*

FIG. 92: *Reconstruction of an Egyptian battering ram, based on a number of tomb paintings of about 2000 B.C.*

As a measure of defense cities were walled. One of the earliest-recorded siege machines was an Egyptian scaling ladder mounted on wheels, although the pharaohs of this era never employed chariots. Since city walls were invariably made of sun-dried brick, the soft material could be prised out to make a breach. The Egyptians devised a portable hut to protect their sappers. This was probably made of a reed or basketwork frame and may, like their shields, have been covered with hides.

FIG. 93: *A sketch of a ship from a fragment of a vase, and a clay model from Crete; about 2000 B.C.*

The development of shipping in the Eastern Mediterranean was quite different from that in Egypt and Mesopotamia. Early pictures of ships from Crete are as frustrating as those from Egypt, but they do show that the vessels were low in the water with a high, upswinging stern.

FIG. 94: *A conjectural reconstruction of a Cretan ship of about 2000 B.C. (below left)*

This reconstruction of an early Cretan vessel is based partly upon pictures of the period and partly upon fishing vessels still being used in Senegal today. The major part of the hull is a hollowed log brought to a projecting point in the bow to help when beaching. The sides of the vessel are built up from a single line of planking, while a curved post and further planks form the stern.

FIG. 95: *Building a fishing boat, East Africa.*

The type of construction suggested for early Cretan
shipping can be seen clearly in this photograph. In the
foreground the inverted hull is being shaped by adze
(compare Figure 85), while the projecting prow, here a
distinct timber, and the additional planking forming the
sides of the vessels can be seen in the boat drawn up on
the beach.

East. As we have said, its contacts must have been entirely by sea,
and it is more than just mildly infuriating that early pictures of
ships recovered from this site are so inadequate that they allow us
to make no reasonable reconstruction. There must by this period
have been some very viable timber-built shipping capable at least
of coastwise traffic, working the northern shores of the Eastern
Mediterranean. But the form and dimensions of these ships escape
us completely, and we shall find that the nature of shipping in this
area will be a recurrent problem for the historian throughout the
period covered by this book.

So far, we have not been able to say a great deal about many

FIG. 96: *Pictogram of a plow from Crete, of about 2000 B.C.,
and a reconstruction based on this and on plows
still in use in western Anatolia.*

> Many aspects of life in Crete can be attributed to her
> contact with countries overseas. For example, the
> Cretans used seals similar to those in Mesopotamia and
> kept records on clay tablets. They developed their own
> script, however. One early pictogram shows a plow, but
> it differs completely from those in use in Egypt and
> Mesopotamia, for it had only a single handle, leaving the
> plowman a free hand with which to goad his oxen. This
> type of plow may well have been an import from Greece
> or Anatolia, but we have no record of equally early
> plows from these areas.

crafts, especially those dealing with perishable materials, but now
with copper and bronze tools, with wall paintings and engravings
and with better-preserved finds from archaeological excavation,
we begin to see just how far many crafts had developed. For example, both in Mesopotamia and in Egypt we now find properly
jointed furniture, although this could probably only have been
afforded by the very wealthy citizens. The ax and chisel both previously existed as flint or stone tools, and we now find them made
of copper and bronze, although as a carpenter's tool it is interesting
to note that the adze seems to have been more popular than the ax.
Wood was cut with a saw which resembled if anything a rather
large kitchen knife with a toothed edge, and from surviving examples it is clear that there was no attempt to set the teeth as with
a steel saw. Often the wood being cut was held in a simple vise
made by driving two posts into the ground and lashing the timber
to be cut between them or into the split of a cleft, upright post.
While smaller holes were drilled with a bow drill, the mortises and

FIG. 97: *The sledge of Queen Shub-ad, from a Mesopotamian tomb; about 2000 B.C.*

FIG. 98: *Egyptian carpenters of about 2500 B.C., as depicted on the walls of a tomb.*

Woodworking now became very sophisticated. Ax and adze were used for rough shaping; timber was lashed between upright supports while being sawed; mortises were cut with mallet and chisel; a final polish was given with blocks of sandstone. Because so little wood has survived from Mesopotamia it might appear that the craft was not so well developed as in Egypt. In fact, rare examples of woodwork, such as this sledge from the tomb of Queen Shub-ad, show that the Mesopotamians were every bit as advanced.

FIG. 99: *Stone bowl from an Egyptian tomb; about 2500* B.C.

carving were done with a wooden-handled chisel struck with a wooden mallet, almost identical in shape to the mason's wooden mallet used to this day.

An enormous industry in stone vases developed in Egypt, and while some of these were made of softer materials, such as calcite, the Egyptians were perfectly able to work granite and other hard rocks. The vases were roughed out in the solid from a block to give a cylinder approximating the shape of the vase. For hollowing out the cylinder a special type of drill had been developed. At the lower end of an upright shaft was fixed a crescent-shaped piece of flint which formed the bit, while the upper end of the shaft was fitted with a crosshandle weighed down by large boulders contained in string bags attached to it, the whole device being rotated by one or two workers. The final surface finish of the vessel was made by polishing with a sandstone rubber, and indeed the same type of sandstone rubber was used by carpenters for finishing off their woodwork.

Like their Mesopotamian counterparts, the Egyptian rulers found it necessary to keep records, but instead of using clay tablets as was the case in Mesopotamia, the Egyptians devised a lighter and more convenient way of storing information in the form of

FIG. 100: *Hollowing stone vases: a scene from the walls of an Egyptian tomb of about 2500 B.C.*

FIG. 101: *A reconstruction of the drill used to hollow stone vases.*

> In Egypt, where there was an ample supply of fine decorative stone, there developed a large industry devoted to producing really superlative stonewares. The vessels were first roughed-out by flaking off the unwanted material from a suitable block, but the final shaping and polishing was done entirely by rubbing with sandstone. The only specialized tool introduced into the industry was the drill for hollowing the stone blocks. The drills were fitted with a crescentic bit of flint turned by hand and weighed down by rocks attached to the handles.

papyrus rolls. The papyrus reed of course grows prolifically in the Nile delta, and we have already seen how it was used for making boats and in the manufacture of rope. Strips of this reed were laid edge to edge on a flat surface and further strips were laid at right angles to them and across them, again, edge to edge. The material was then beaten hard with a heavy wooden mallet, causing the strips to fuse and to become flattened, thus forming a paperlike material that could be rolled up for storage. This was, of course,

FIG. 102: *Two men beating out papyrus: from an Egyptian tomb of about 2500 B.C.*

Unlike the Mesopotamians, the Egyptians seldom kept their records on clay tablets. Instead they wrote in ink on papyrus. To make this material strips of the papyrus reed were laid on a flat surface and further strips were laid across them at right angles, and the whole was then hammered until all the strips welded to form a continuous sheet. For storage the papyrus was rolled in convenient lengths.

FIG. 103: *Two early pictures of looms from a Mesopotamian seal and an Egyptian painted dish, both before 3000 B.C., with a reconstruction of this type of loom.*

FIG. 104: *Horizontal looms from an Egyptian tomb painting; about 2000 B.C.*

FIG. 105: *A horizontal loom used by Andean peasants today.*

The earliest pictures of looms appear shortly after 3500 B.C. on a seal from Mesopotamia and a painted dish from Egypt. Both were essentially of the same type: the warp threads were stretched between two beams which were lashed to short upright posts at no great distance from the ground. The Mesopotamian seal gives no impression as to how the weaving was done, but the Egyptian dish shows three rods lying across the warp. This may mean that three different lines of weave were being produced. In other words, a step may have been taken toward making pattern weaves, rather than plain weave, on the loom.

Pictures of looms of the period 3000–2000 B.C. from Egypt show the same type of loom in use, although as an artistic convention they are invariably shown in false perspective as though rotated through a right angle. Looms of this type are still in use today in many remote parts of the world.

FIG. 106: *Beermaking in Egypt: a scene from a tomb of about 2500 B.C.*

an easy material on which to write with ink, and although the Egyptians started their system of recording with pictograms, as did the Mesopotamians, there was no real necessity drastically to modify this method of writing. Thus their pictograms slowly changed their meaning and became totally stylized to give the familiar hieroglyphs; but throughout the history of ancient Egypt there was to be no other great change in methods of writing and recording, except that very early the Egyptians reduced their hieroglyphs to a highly simplified cursive form that could be written rapidly by hand.

Cloth continued to be made on a very simple form of loom in which the warp threads were stretched between two posts, held a short distance horizontally from the ground by forked pegs. Into this warp was woven the weft that had been wrapped round a bobbin. For their finer raiment the Egyptians used linen—the flax plant being found native in the Nile Valley. Wool, on the other hand, was not much favored by the Egyptians, although they did make cloaks of the material. But farther north, particularly in Anatolia, such little evidence as we have would suggest that nearly all cloth of this period was made of wool.

The date palm and barley provide us with the first direct evidence we have that both the Mesopotamians and the Egyptians were making fermented drinks. Date palms are prolific in both areas, the fruit itself is rich in sugar, and in these warm climates fermentation is fairly rapid. Thus, while the first proof of the making of date wine comes from this period, there is little doubt that it was prepared by the people in these areas long before. Since the fruit itself normally contains the micro-organisms necessary to cause fermentation, the making of date wine was not of itself a very complex process and fundamentally one only had to have a large jar in which to put the date mash while fermenting and some device with which to strain it once fermentation took place. The making of beer, on the other hand, was a far more sophisticated process, for some of the cereal, usually barley, first had to be allowed to germinate; in so doing much of the starch in the grain was converted into sugar, making the ultimate drink both palatable and slightly sweet. Hence, the first stage was to moisten the cereal and allow it to germinate for a short time. The malted cereal was then made into loaves and baked slightly. These malted loaves were then crumbled and added to water with more grain, and this mash was allowed to ferment for three or four days. It was then strained, bottled, and stoppered to prevent the further fermentation which would make the drink acid.

At this period, too, the grape was also being cultivated and the fruit converted into wine, and apart from scenes depicting the making of wine we also have records of vineyards. As far as

FIG. 107: *Drinking beer through straws as depicted on a Mesopotamian seal of about 2500 B.C.*

Beer was being brewed in both Egypt and Mesopotamia. Malted grain was first made into loaves, which were then allowed to ferment in water. The beer was subsequently strained, bottled and stoppered. The product was presumably of rather doubtful quality since the Mesopotamians invariably depict themselves drinking it through straws!

FIG. 108: *Early Egyptian grape presses from a tomb painting; about 2000 B.C.*

one can gather, these vineyards were the property of the rulers both in Mesopotamia and in Egypt. It seems unlikely that the common man could ever hope to drink wine made from the grape. It has even been suggested that grape wine was reserved entirely for the gods. But the actual process followed seems to be more or less the same that we know today. The grapes were harvested, trodden, allowed to ferment; the juice was then strained and bottled. Sometimes, too, grape or date wines, and even the beer, were flavored with herbs. In many records we read of more than one variety of date wine and more than one variety of beer, and it

FIG. 109: *A reconstruction of the Egyptian bag press.*

Both the date palm and the grape were used for making wine. To extract the juice from grapes a simple press had been devised. The grapes were put in a tall linen bag which was then stretched between two posts. Bar handles were placed at both ends of the bag and twisted.

FIG. 110: *Egyptian gardening: a scene from a tomb of about 2000 B.C.*

Many roots, vegetables and shrubs other than the vine were now brought into cultivation, and the various rulers appear to have delighted in rearing exotic plants. In this scene the two men on the left are gathering fruit while the man on the extreme right is planting out seedlings. Two other men carry water from a canal or reservoir.

would seem that the additions made to these drinks and the precise methods of manufacture were jealously guarded secrets— perhaps the first manufacturers' secrets in this history of technology.

Apart from grapes, other fruits and some vegetables were being brought into cultivation, and the monarchs both in Mesopotamia and Egypt seem to have prided themselves on the variety of plants growing in their gardens, while considerable effort appears to have gone into the establishment of exotic fruits and vegetables. Naturally these gardens required watering all year round, and although we may be sure that in Mesopotamia perennial irrigation had already advanced considerably by now, in both Mesopotamia and Egypt we have direct evidence of a water-lifting device, the shaduf. Essentially this was a beam pivoted into the top of an upright post. At one end of the beam was attached a rope on the end of which was the vessel to contain the water; on the other end of the beam was a counterweight. The vessel was lowered by hand into the water of the river, raised, emptied into either a cistern or into an irrigation channel and then lowered again into the water. The shaduf, of course, took a great deal of the backache out of raising water, which before this time presumably had to be done entirely by hand. Nevertheless, it remained a

FIG. 111: *The impression of a Mesopotamian seal showing a shaduf; about 2000 B.C.*

slow and tedious job; it was many centuries before a device was invented capable of doing this work in which a draught animal provided the power.

By the year 2500 B.C. an Egyptian pharaoh caused to be built the largest of all the pyramids, the Great Pyramid, the base of which is 756 feet square. It has been estimated that a total of not far short of 2,500,000 blocks of stone were required to build it and that the average weight of each block was some two and a half tons. It is extremely difficult to come to any reasonable estimate of what this means in terms of human labor, but more than one authority believe that a working force of some 100,000 men, working during the season of flood—that is to say, from the end of July to the end of October—were required to transport and lift the stone, and that this operation took in all some twenty years. Of course, at the quarry and at the pyramid there would have been permanent gangs of workmen quarrying and trimming the stone blocks throughout the whole year, but these gangs would not necessarily have been by comparison very large—perhaps a few thousand men at each location. Such structures were, of course, gigantic undertakings demanding a considerable amount of planning, mathematics, practical geometry, and the basic elements of engineering. But whatever the driving force that caused them to be built and however wasteful of manpower they may seem to us today, the pyramids are only one lasting symbol of a period of

FIG. 112: *A shaduf depicted on an Egyptian tomb painting; about 1500 B.C.*

FIG. 113:
The shaduf in use in Upper Egypt today.

Perennial cultivation meant that plants had to be provided with water. Sometimes water was brought from the river or reservoirs by hand, but the first illustrations of a water-raising machine, the shaduf, belong to this period. A container was suspended from the end of a beam with a counterweight. The shaduf obviously increased the volume of water that a single workman could raise.

FIG. 114:

Diagram showing the evolution of the pyramid.

In Egypt the period 3000–2000 B.C. was the great era of pyramid building. Their origin may be traced to the "houses of the dead" (*mastabas*) in which, to prevent robbery, the dead were often buried in a chamber beneath the floor. In the "Step Pyramid" (2800 B.C.) the *mastaba* was added to in two major stages. The pyramid of Seneferu at Meidum (2700 B.C.) still retained the *mastaba* and step structure, but was given a final smooth facing of limestone. Finally, in the apogee of all, the Great Pyramid of Cheops, the *mastaba* and steps were eliminated and the pyramid took on the familiar form.

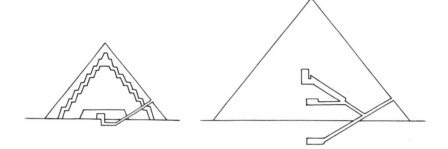

great enterprise, for the Egyptians now were beginning to look much farther afield for the raw materials they needed for their industries. Regular armed convoys of miners, for example, were being dispatched into Sinai, there to quarry the copper ores, while regular expeditions were being mounted to go farther and farther up the Nile in search of gold, ivory and other precious materials and even the timber required for their buildings.

Good timber, never plentiful in Egypt, was then nonexistent and timber that was not brought down the Nile had to be sought from other countries of the Eastern Mediterranean. The Egyptians were obliged to build seagoing craft of wood. Their method of constructing these ships, however, betrayed their origin, for in fact they were little more than overgrown papyrus boats made of planking in which the edges of the planks were quite literally sewn together. They thus lacked either keel or ribs. Added strength was given athwartships by running a deck planking through mortises in the gunwale, while to prevent the vessel's breaking its back a large stay was carried from stem to stern. Raised above the deck by forked timbers, this stay was maintained taut by a tourniquet passed through the rope and lashed to one of the uprights. Such a vessel could carry nothing in its hold, and passengers and cargo had to travel on the deck as, too, did the crew, who now appear to have abandoned the paddle and taken to the oar. One is left wondering whether or not the Egyp-

FIG. 115: *The Gisa group of pyramids.*

The Great Pyramid, seen here in the foreground, was a gigantic undertaking by any standards. It is 756 feet at the base and rises to 482 feet; it has been estimated that its construction took 2,500,000 blocks of stone of an average weight of two and a half tons apiece. With a possible force of 100,000 men it may have taken twenty years to build.

FIG. 116: *Egyptian wooden seagoing ship of about 2500* B.C.:
a model reconstruction.

Wooden seagoing ships were fitted with a bipod
mast, but in order to prevent them breaking their
backs at sea a rope was carried from stem to stern,
held above the level of the deck by forked wooden
posts. This rope was kept taut by a tourniquet. The
crew, who now invariably rowed rather than paddled
the ship, sat on stools on the deck. The cargo, too,
had to be carried on deck since the hull was too
frail to support the load.

FIG. 117: *A large rowed riverboat: a model from an Egyptian tomb of about 2000 B.C.*

The over-all deck and individual seats for the oarsmen can clearly be seen in the tomb model. The steering paddle was lashed to a vertical post, and passengers were carried on deck.

tians borrowed the idea of rowing from other maritime nations of the Eastern Mediterranean, and furthermore one must doubt the efficiency of the crews, for in their earliest forms we find the oarsmen of these ships seated not on thwart planks but on separate stools.

While in Egypt copper remained the normal toolmaking metal, in Mesopotamia a subtle change was taking place in the field of metallurgy, for the smiths had found that a greater control of the product could be achieved by adding tin ore to metallic copper to produce their bronze, rather than by smelting the tin and copper ores together as they had done hitherto. Their products now became more and more standardized and the percentage of tin present in their bronzes remains consistently around eight per cent—a very good figure for the production of a hard but non-brittle metal. It nevertheless remains something of a mystery as to where all the tinstone came from, and one suspects that the Mesopotamians were sending prospectors far afield looking for their raw materials. At this time too the Mesopotamians had discovered that many of the deposits of sulphide ores of lead—galena—also contained considerable quantities of silver, and they had devised a means of recovering the silver from this source. The galena was placed in a large roasting furnace with the result that the lead was either volatilized or absorbed by a thick bed of ash, while the silver, initially present as the sulphide, was reduced to the metal and could be recovered from the ash by washing. But it will be noticed that at this stage the lead was not utilized, but was allowed to go to waste.

Shortly after the year 2300 B.C. we find the whole of Mesopotamia unified under a single Akkadian ruler, Sargon the First, who pushed his frontiers up into Syria and thus ruled over a small empire extending from the Eastern Mediterranean to the Red Sea. Like his Egyptian counterparts there can be little doubt that Sargon's chief concern in this enterprise was to attempt to control the sources of supply of his raw materials. One is tempted, therefore, to suppose that much of the tin required for bronzemaking came from the mountains of Syria and of eastern Turkey, while Syria could also of course have provided timber and a trading outlet to the Eastern Mediterranean.

Thus, for a brief period we find the two most technologically advanced areas of the world ruled by two supreme monarchs. A scribe of a later period, however, records that one of Sargon's successors was forced to carry out a campaign to quell an uprising on his northern frontier. The attackers were a coalition of seventeen princes who ruled over northern Syria and eastern Turkey, and among these princes was one whose name betrays him to have been of Indo-European origin—that is to say, a member of a linguistic group whose homeland lay in a more northern area of Asia. If the scribe's account is true, then the Akkadians were already in contact with the people who were to have the most profound effect on the technology of the early Near East, for it was the Indo-Europeans who came from the steppes of Asia who, almost certainly, first domesticated the horse.

Before we move on to the next great epoch, however, we must say something more about both Egypt and Syria, for many small technological innovations appear during this period that show considerable sophistication.

The old glazed soapstone ornaments, the so-called Egyptian faïence, were now replaced in both areas by a completely synthetic material. White sand was mixed with natron, a naturally occurring form of sodium carbonate, shaped and heated so that the whole mass fused. It was to this synthetic core that the blue glaze was applied. The first step, in fact, had been taken toward the making of true glass, for the fusion of quartz and soda with the addition of a little lime to make it stable remains the basis of much of our glass even today. At this early period it is clear from Mesopotamian texts that the glassmakers were unaware of the need to add lime to ensure a stable glass. Fortunately for them, however, their raw materials already contained sufficient lime without the need for its deliberate addition. The synthetic core made of sand and soda if overheated would, of course, have become molten, and there exist many examples of this type of ornament in which the heating had been stopped only just before the object melted completely and lost its shape. So there is some justification in supposing that the discovery of glass resulted from observing accidents in which faïence was heated until it melted. Just before 2000 B.C. we find in Mesopotamia the first appearance of true glasses. But the objects themselves betray the fact that the Mesopotamians had still not appreciated fully the potentialities of their new material, for instead of molding the material while it was hot, as one might expect, it was worked in the cold and was largely cut and polished with abrasives in the same way that these early craftsmen dealt with harder decorative stones. Nevertheless it is evident that a great deal of experimentation was being carried on, for at this time we also find the appearance of a limited amount of lead in the glazes on faïence objects. The effect of lead in a glaze, in a glass, is to give it a greater brilliance, and while we have no idea how the discovery was made it is clear that the craftsmen who made faïence were on the lookout for better ways of doing things.

The same spirit surely prompted the introduction of what we might reasonably call a gadget which was attached to the plow. Earlier we saw that the Mesopotamian plow started life very like the Egyptian one: as a pronged stick dragged through the ground, and to which a sole was later attached to make a more effective furrow. Now, in the forepart of the sole we find a vertical hole had been drilled through it, and into this hole had been inserted a vertical tube, the top of which was shaped like a funnel. This curious device was in fact a seed drill, and there exist a few illustrations of

FIG. 118: *A model of a plow team from an Egyptian tomb of about 2000 B.C.*

the sower pouring seed into the funnel. This gadget, which certainly ensured that all the seed ended up in the furrow, had a fairly long life, for nearly fifteen hundred years later we still find the same type of seed-drill attachment illustrated on the tile walls of a royal palace in Syria, even though by this time the drill may have become an archaism.

Prior to Sargon's conquests, the temple, although being the main building of any city in Mesopotamia, had not dominated the scene as it did from now onward. Early temples had often been built on a low raised platform of mud brick, but now it became the habit to build the temple on the top of a series of stepped platforms so raising the temple far above the level of the remainder of the city. The building of these stepped platforms, or ziggurats, presented certain problems, for had they been constructed entirely of mud bricks, due to the sheer volume of the building they would have been unstable, while it was also beyond the economic resources of these people to have created ziggurats entirely of fired brick. Thus, in order to bind the whole structure together and prevent excessive movement, layers of reed matting were often

FIG. 119: *Part of the impression from a seal from Mesopotamia showing a plowing scene; about 2000 B.C.*

FIG. 120: *Symbols depicting plows from Mesopotamian seals (3000–2000 B.C.), and a reconstruction of this type of plow based on these pictures and on plows still in use in Iraq today.*

In Egypt plows remained virtually unaltered in design throughout this period, but in Mesopotamia a sole, as a separate piece of timber, was added, and later a seed drill was inserted into the sole. In the seal from the very end of this period one can see the sower pouring the seed into the drill while plowing is in progress, but the plow alone often appears as a symbol on much earlier seals.

incorporated between the layers of mud brick—fired brick and stone being reserved only for facing.

The building of pyramids in Egypt and of ziggurats in Mesopotamia demanded not only some knowledge of geometry but also that in each area there existed a uniform method of measurement. This of course is not meant to imply that uniform systems of measurement were devised simply in order that pyramids and ziggurats could be built. Indeed, long before the Egyptians ever thought of building a pyramid they had needed to make an annual survey of agricultural land after each flood, and perhaps because of this the Egyptians became adept at surveying; even as early as 3000 B.C. we find that standards of weight and of length were being maintained in temples and royal palaces of both Egypt and Mesopotamia.

Early units of length were based upon measurements that

FIG. 121: *The Ziggurat, Ur, during excavation.*

FIG. 122: *A reconstruction of Ur Ziggurat as it may have looked in antiquity; about 2000 B.C.*

The ziggurat became the centerpiece of most Mesopotamian cities. It was essentially a stepped platform on top of which was built the temple. Ziggurats were built almost entirely of sun-dried brick, and in order to support the superimposed load of the building without the brickwork crumbling, layers of reed matting were incorporated as the construction progressed. The matting served as a reinforcement in much the same way as do steel rods in a modern concrete building.

could be made from one point to another on the human body. Thus throughout the Near East we find the cubit, from the point of the elbow to the tip of the middle finger; the span, from the tip of the little finger to the tip of the thumb of an outstretched hand; the palm, usually measured across the knuckles; the finger width; and the foot. Furthermore, the smaller units—spans, palms, finger widths and feet—were generally related as subdivisions of the cubit. But in Mesopotamia and Egypt, very early, a totally different view developed as to how many subdivisions there should be to each cubit as well as what the subdivisions should be called. Thus the Egyptian royal cubit was divided into seven palms, each palm being reckoned as four finger widths, giving a total of twenty-eight finger widths to the cubit. Elsewhere in the Near East we commonly find the cubit divided into two feet, the feet into three palms,

FIG. 123: *Diagram showing how Egyptian and Mesopotamian measures of length were based upon parts of the body.*

Elaborate buildings, and other construction work, demanded the use of standard units of linear measurement. The cubit, from the elbow to the tip of the middle finger, was in general use throughout the Near East. In Mesopotamia the cubit was further subdivided into feet, hands and fingerwidths (inches) and was, hence, ancestral to our own system of measurement. In Egypt the cubit was divided into seven hands, each of four finger widths. Needless to say, the cubit was not of identical length in all regions.

and the three palms again subdivided into four finger widths—a system that can be readily recognized as ancestral to our own measurement of feet and inches.

Since both Egypt and Mesopotamia had now become commercial countries in the strictest sense of the word, not only did they have to have standard units of length but also standard units of weight in order to regularize their transactions. Initially balances were used only for weighing precious materials and the earliest units were hence small, the major unit being the shekel, itself divided into a number of grains. Thus, while the grain itself was the hypothetical weight of a grain of corn, the shekel might vary from 120 grains to over 200. In time and as the need arose, larger units —multiples of the shekel—were devised: the mina reckoned at anything from twenty-five to sixty shekels, and later the talent reck-

FIG. 124: *Egyptian stone weights from Sinai; about* 2000 B.C.

FIG. 125: *Egyptian balance as depicted on wall paintings; about* 2000 B.C.

Trade in metal and other precious materials equally demanded standards of weight. These were all based upon the theoretical unit, the weight of a grain of wheat, and different people used various multiples of the grain as larger units. These varied very considerably from one country to another, so that a merchant trading in the Mediterranean, for example, would have to carry a set of weights appropriate to each country he visited. The Egyptian weights shown in this instance were those used to measure the dried-fish ration of miners extracting copper ores in Sinai.

To judge by the accuracy of the weights themselves, balances must have been fairly sensitive. Simple two-pan balances were often depicted, but never in sufficient detail to show how they were pivoted or adjusted.

oned at sixty minas. While, however, the mina and talent were in use throughout most of the Near East, in Egypt a system of metric weights was in use, which seems to us today strangely ahead of its time.

The actual weights themselves were normally made of some hard polished stone with the denomination of the weight cut into it. In Mesopotamia, for reasons that have never been explained, these weights usually took the form of a duck preening itself, but Egyptian weights tend to be more prosaic and are usually geometrically shaped blocks of stone with rounded corners and edges. As with measurements of lengths, units of weight have been found as standards in palaces and temples, but here again there is a very considerable variation in the value of the standards from area to area, and even from city to city. The balances were simple—equal armed scales with a pair of pans, and although we know of the balance as a symbol in early writing and from many illustrations, there is little to suggest at this period how and of what the fulcrum was made; hence it is very difficult to get an estimate of the accuracy of weighing that was possible with these early balances.

It is clear both from their achievements and from existing records that the Egyptians and the Mesopotamians during the one thousand years just reviewed in this chapter had made considerable advances in the mathematical field. The Mesopotamians, for example, for no apparent practical purpose, had already learned how to resolve simultaneous equations, while the Egyptians, through a study based partly on the annual flooding of the Nile and partly on the behavior of the heavenly bodies, had established a reasonably accurate calendar. From their experience in surveying the land, the Egyptians had also discovered how best to lay out a right-angled triangle—and indeed had found their own solutions to the knotty problem of the square of the hypotenuse—using a system in which the hypotenuse was divided into the same number of subdivisions as the two equal sides of a right-angled triangle. This simple ruse allowed them to multiply and divide areas as well as lengths.

Few if any of the technologies so far considered require in their working any knowledge of mathematics at all, but in a somewhat roundabout way the ability to weigh accurately must have helped the metallurgist in more than just his commercial transactions. Thus in Mesopotamia we find an almost set formula for an operation performed by the goldsmith. The inscriptions usually read something like this: "x minas of gold were put into the furnace and after heating y minas of gold remained; the loss as a result of heating equals x minus y minas." Clearly the idea must have occurred to the Mesopotamian goldsmiths that gold could be obtained so pure that when put into the furnace it would suffer no further loss, and from this time on pure gold became the primary standard of exchange.

5/Of Chariots, Sea-going Ships and the Expansion of Trade

(2000–1000 B.C.)

THE POLICIES of Egypt and Mesopotamia of pushing out their frontiers to include areas rich in mineral and other resources had their dangers. In modern parlance, apparently the administration in both areas was extended beyond its capabilities. This was serious enough, but the northern boundaries of Mesopotamia brought the people of the Euphrates valley into contact with groups of tribesmen living in the hilly areas of eastern Turkey and northern Syria and by 2000 B.C. an infiltration into this hill area of other people from more northerly regions of Asia had begun to take place. As we shall see later, some aspects of Mesopotamian civilization had already been diffused by way of the same hill areas into the Asiatic steppe and by this period the nomadic herdsmen had already acquired the wheeled vehicle, almost identical in its cumbersome shape and method of manufacture to the early carts of Mesopotamia, if indeed they were not its original inventors. But instead of the ox, which was too slow, or the onager, which was not strong enough, the steppe dwellers had apparently learned how to tame and harness the horse. The newcomers into the mountainous area north of Mesopotamia had thus on the one hand domesticated horses, which they either brought with them or acquired from their northerly neighbors, while on the other hand from the Euphrates valley they had learned sophisticated methods of working wood that so far had been applied only to the manufacture of such things as furniture. Within a very short space of time these hill dwellers had discovered how to build a light maneuverable vehicle of war that could be drawn by a team of horses at great speed. The chariot was now to loom very large in the affairs of man.

This vehicle demanded a number of innovations. In the first place the horse could not be controlled either by a simple halter or by a ring through its nose; instead, the bridle and bit had to be used. The yoke and pole were left virtually unaltered, but the chassis and wheels were completely redesigned. The chassis now appeared as a light wooden frame designed to do no more than carry the two or three warriors, while the wheels were made of a hub from which radiated four spokes set into a felloe of bent wood. Considering the stresses and strains involved in driving a wheel hard over rough ground, one appreciates that hub, spokes and felloe would probably each have had to be made of a distinct type of timber, and that a wood suitable for one purpose might well not have served for the others. Although the only surviving chariots come from Egyptian tombs of a rather later date, it is equally clear that the wheelwrights were fully aware of the need to pick the right timber for the job: a hard, dense wood, little prone to cracking, for the hub; a straight, tough, inflexible timber for the spokes; and a straight-grained, easily bent timber for the felloe. Elm, oak and ash, still used in Europe today for the making of cartwheels, were the same timbers selected by the makers of chariots.

Obviously the chariot was a weapon that could and did revolutionize warfare. The speed of an army was no longer that of a

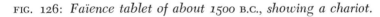

FIG. 126: *Faïence tablet of about 1500 B.C., showing a chariot.*

FIG. 127: *Reconstruction of an early Egyptian chariot, based upon tomb paintings and on two surviving vehicles.*

FIG. 128: *Diagram showing how the yoke was adapted for use with horses, in its simplest form (above) and in the more elaborate Egyptian ogee-curved form of about 1500 B.C. (below)*

The earliest true chariots were lightly constructed, highly maneuverable vehicles carrying a crew of a driver and one or two warriors. Their speed, relatively much greater than that of infantry, meant that military commanders were forced to adopt totally new tactics. The chariots were drawn by horses yoked in pairs and controlled by bridle and bit. Two Y-shaped pieces of wood were attached to the yoke, so as to fit the narrow shoulders of the horses. Early chariot wheels had only four spokes, which must have limited the terrain over which they could travel.

marching column of infantrymen, although of course infantry remained the major part and backbone of any army. But the chariot
could operate ahead of, and at a considerable distance from, the
main force. Any army commander careless enough to allow his
column to straggle ran the almost certain risk of seeing it decimated
piecemeal by his opponent's chariot force. From now on the military commander had the means of carrying out the outflanking
maneuver in the grandest possible style.

The political results of the introduction of chariotry are fairly
self-evident. It need hardly be said that new regimes appeared
both in Mesopotamia and ultimately in Egypt. The chariot, however, had more lasting effects and its introduction was to go a long
way toward changing many basic technologies. All the evidence
suggests, for example, that the people of the Mesopotamian valley
were now cut off from their main supplies of tin and that they
began to seek other sources in the west by way of the Levant, thus
bringing them even more into commercial contact with the people
of the islands of the Eastern Mediterranean, that is to say Cyprus,
Crete and those of the Aegean, as well as with people dwelling on
the Anatolian coast.

The fragmentary evidence that we have, largely representations on cylinder seals, a few clay models and sketches that can only
be referred to as doodles, shows that by now at least two types of
vessel were being used in the eastern end of the Mediterranean.
The one was probably a broad-beamed vessel with a high, sweeping bow and stern which at first sight appears to be simply an overgrown version of the reed boat. However, even if this were the
ultimate source of inspiration the vessel must have been plank-
built. The second type of vessel, although equally high-sterned,
was apparently less broad in the beam with a low bow rising
vertically from a projecting "ram." This vessel looks as though it
were essentially an extremely large tree trunk that had been hollowed and shaped and to which washboards and a high stern had
been added. This is a type of construction which, without the high
stern, can still be seen off the coast of Senegal today, where it is
used by fishermen who, it should be noted, are quite happy to sail
the vessel despite its lack of keel.

Both types of craft were propelled by oars or by a square sail.

FIG. 129: *The impression of two seals from Crete; about 1500 B.C.*

FIG. 130: *Reconstruction of a ship from the Eastern Mediterranean, based upon two badly damaged Egyptian tomb paintings of about 1500 B.C.*

FIG. 131: *Conjectural reconstruction of this type of merchant ship.*

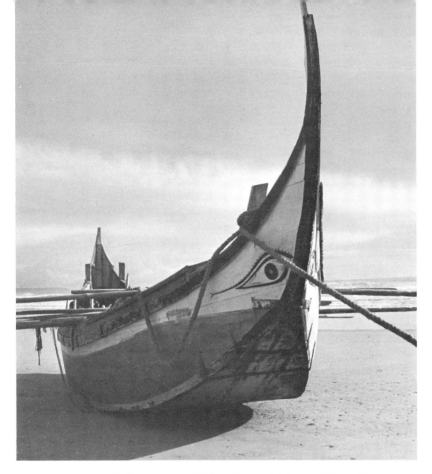

FIG. 132: *A Portuguese fishing boat which possibly echoes some of the features of Mediterranean ships of this period.*

Merchant ships from Crete and the Levant that traded with Egypt were apparently short, broad-beamed vessels, rising at both stem and stern. Cretan seals give only the vaguest impression of what they were like, and the reconstruction is based very largely on badly defaced paintings from two Egyptian tombs. The vessels appear to have been decked at a level a little below the gunwale, and it was upon this that the cargo was carried, protected to some degree by hurdlework rising above the gunwale. The square sails differed from those of the Egyptians in that they were attached to the yardarm only in the center and at either extremity. It is far from clear whether the hull was strengthened by internal ribs, but the general form of the hull seems to have been preserved in small lighters still used in some Black Sea ports and in fishing vessels used by the Portuguese.

Apparently the vessels were rowed by five to ten oarsmen on each side, although the smaller number is the more common, and on this basis we can guess that the ships ranged from perhaps thirty to fifty feet in length. Oars were, of course, essential when the ship was becalmed or when working into wind, but when the wind lay abaft a square sail was used, hoisted to a single mast stepped amidships, a practice very different from that used hitherto in Egypt. It is impossible to say whether or not these vessels carried a deck, but on the face of things it seems unlikely. Some representations would suggest that the washboards were attached to the hull by means of a series of vertical posts, and if this is so then the idea of internal, strengthening ribs was already present in embryo. The high stern to protect the helmsman, and also presumably the master, from the seas when there was a following wind is a feature that was to be retained in much Mediterranean shipping from this time on and is still to be seen in many local types of craft used in that area today.

It is one thing to be able to say that the people of Crete and the Levant had vessels adequate for sailing on the Mediterranean: it is quite a different matter to say with any certainty precisely where they went and in what commodities they were dealing. However, even as early as 2500 B.C. it is clear that a number of small colonies of people coming either from the eastern end of the Mediterranean, or more probably from Malta or Sicily, were established in southern Spain and Portugal. These people, living in small fortified cities and using bronze for toolmaking, were surrounded by an indigenous population who still relied on stone tools and weapons. The area is rich both in copper and tin so that one suspects that here at least was one source from which the people of the Eastern Mediterranean were deriving the metals so essential to their technology.

Among the countries to profit from the presence of these early merchant venturers was Egypt, where we now find that bronze rather than copper had become the general metal for toolmaking. From this period, too, we begin to find small blue beads made of Egyptian faïence spread widely among the scattered and backward people of Western Europe, which allows us to think that quite

FIG. 133: *Drawing from part of an Egyptian tomb painting of about 1500 B.C., showing foundrymen casting bronze doors.*

The foundrymen are casting a bronze door, and the metal is being poured into a series of cups leading into the large clay mold. A previously cast door can be seen above and to the right of the foundrymen. In this scene from an Egyptian tomb the men working the bellows must be envisaged as rocking from side to side as they tread down first on one drum to expel air, and then on the other. The function of the string attached to the skin of the bellows (upper left) is to raise the skin, so filling the drum for the next tread.

FIG. 134: *Diagram showing how the crucible was probably handled.*

FIG. 135: *Drum bellows in use until quite recently in Rhodesia.*

With the introduction of the bellows for raising the temperature of the furnaces, it became possible to heat larger volumes of bronze than when the fire had to be blown by lung. The bellows were either whole skins or skin-covered drums worked in pairs.

a considerable trade had been built up along the length of the Mediterranean.

Insofar as we can determine, by 2000 B.C. tinstone was still being added to metallic copper whenever it was required to make bronze, so that it is highly likely what was being imported into the eastern end of the Mediterranean was either the tinstone itself or more probably the already formed alloy. This does not mean, nevertheless, that no further progress had been made in the Eastern Mediterranean in the field of metallurgy. During the next five hundred years the whole process of bronzemaking became extremely sophisticated. Shortly after 2000 B.C. we know for certain that the bellows had been invented and was being used as a means of raising the furnace temperature, for we find an inscription on one of the Mesopotamian clay tablets which is a demand for the skins of two large he-goats for the making of bellows for a bronze-founder, while at a slightly later date an Egyptian tomb painting shows us twin bellows in use being worked by foot. The bellows must have allowed a far larger scale of production than did the blowpipe, permitting not only the production of more tools and weapons but also, because greater quantities of the metal could be melted at the same time, the making of larger castings. A tomb painting in Egypt illustrating the casting of a large bronze door shows us that the mold had a whole series of sprue cups through which the molten metal was poured, a situation that would have been quite impossible if all the metal had had to be made fluid by the older expedient of raising the furnace temperature by means of mouth-blown pipes.

Chemical analyses of bronze implements show us that up to this point metallurgists had had to be highly selective in the copper ores they used. They were unable to use ores containing sulphur because it formed a gas within the molten metal which, on cooling, became porous and unserviceable. It was now discovered, however, that by roasting the sulphide ores in an open fire the bulk of the sulphur could be eliminated and the ores could then be reduced in a normal furnace to provide a sound metal. Since sulphide ores are, in fact, of far more common occurrence than other types of copper ore, this discovery made a much larger quantity of metal immediately available.

Finally, the old practice of adding tinstone to metallic copper gave way to a different method of alloying: the tinstone was first reduced to metallic tin in a furnace and the two metals were then heated together to form bronze. This system allowed complete control over the relative proportions of the two metals being used, and from now on bronzes with low and with high tin content were being used for specific purposes. Thus, for the making of mirrors a very high proportion of tin was added, which while making the metal far whiter also made it very brittle and thus useless for tools. For tools and weapons a proportion of about eight per cent of tin was retained, but where, for example, a bone handle had to be fitted to a bronze knife, the rivets were often made of an alloy containing even less tin so that they would be more easily hammered without becoming brittle since they could not be annealed without damaging the bone handle.

Bronze now became so common that it was in everyday domestic use throughout the whole of the Near East, and stone tools of any kind became increasingly rare. Nevertheless, despite the large quantity of bronze available, from about 2000 B.C. onward iron tools, weapons and even ornaments were occasionally produced. Iron does not normally occur in nature as the metal but usually as an ore, although metallic iron is to be found in very small quantities in some meteorites. Thus while some of the very earliest iron objects may be seen as having been manufactured from meteoric iron it is unlikely that this applied to most of them, and it is fairly evident that over a long period, perhaps some seven hundred years, metallurgists were experimenting with iron, never quite managing to master the techniques required in its production.

Unlike any of the metals that were so far known—gold, silver, copper, and tin—iron could not be made fluid in the furnaces used by these early metallurgists, for the temperature was always too low. Even so, if the conditions were right, the ores of iron could be reduced to the metal, but the end product as it was removed from the furnace was a hard spongy mass that looked very unlike a serviceable metal. To make anything of it, it had to be repeatedly reheated and hammered while red hot in order to forge it into a bar of solid metal. This process was quite unlike any metallurgical technique so far known.

Unfortunately, we know remarkably little about the discovery of ironworking, and although it is easy to suggest that in the first instance a very pure red-oxide ore, red ocher, was reduced in small quantities and was heated and hammered as one might hammer and anneal bronze, we still know next to nothing about how and where the full-scale production of iron really started. Historically, the first people that we know of who had a limited but steady supply of iron were the Hittites, whose kingdom occupied very approximately the eastern half of modern Turkey. An existing letter from a Hittite ruler to an Egyptian pharaoh tactfully explains that he cannot let him have iron swords at the moment as good iron is not to be had. In short, we are not even certain that the Hittites manufactured their own iron, and they may well have acquired it from a neighbor. Many authors today suggest that large-scale iron-working may have started within a region roughly defined by the sides of a triangle running from the south of the Caspian Sea to north Syria to the southeastern end of the Black Sea, but concrete evidence of this is still lacking.

Although the ores of iron are far more widespread than those of copper, for indeed there are very few countries that can produce no iron ores at all, the technique of winning the metal was so different from that of copper production that the spread of the knowledge of ironworking appears to have taken a very considerable time. Not only did the craftsman have to learn to forge his metal once it had come from the furnace, but he also had to learn to shape it while still red hot and to weld one piece of metal to another by heating them both to red heat and then hammering them until they welded. Furthermore, it was necessary to exercise a far greater control over the atmospheric conditions in the furnace than was necessary during the smelting of copper, for the iron ores, never becoming molten, had to be converted to the metal by carbon monoxide gas in the furnace, which was not essentially the case with copper ores. It would have been impossible to produce any quantity of iron at all without bellows, or some other system of creating a steady and controlled forced draught, and, although in Mesopotamia the bellows first appears, as we have seen as a bronze-founder's tool it is far from certain that the idea was not borrowed from people of the north already working small quantities of iron.

In order to work the red-hot metal, tongs, a heavy iron hammer with a long handle and a substantial anvil were essential, whereas all the hammering processes with copper and bronze could have been, and in fact were, carried out in the cold, thus avoiding the need for forging equipment.

It has been suggested that ironworking remained the monopoly of the Hittites and their immediate neighbors over a long period of time, and that it was only with the fall of the Hittite empire that ironworking spread to other parts of the Near East. Even if this hypothesis is correct, the metal was used only for the making of weapons, and although from 1500 B.C. onward iron weapons became increasingly more common it was not until the last century before 1000 B.C. that we find the metal in use throughout the larger part of the Near East—even then it was still not being produced in Egypt. As we shall see, shortly after 1000 B.C. the technology of ironworking spread more rapidly. Then, and only then, do we begin to find tools rather than weapons made of this metal. Thus, for the whole of the period being considered in this chapter we must remember that bronze was the universal metal for all tools.

Apart from the improved techniques in working tin and copper, from about 2000 B.C. onward in both Egypt and Mesopotamia the making of glass became increasingly more sophisticated, a large part of the production going into the making of glass vessels and beads. Small glass bottles were usually made either by dipping a friable core of sand and some organic adhesive into a crucible of molten glass, or less probably by covering the surface of the core with broken and finely ground glass material and then fusing it in an oven. In both cases the core was extracted at the end of the operation. Mesopotamian texts make it quite clear that they had mastered the technique of producing quite a wide range of colors in their glasses. Copper ores were used to give turquoise blues, cobalt ores to produce darker blues, iron ores to provide yellows, and tinstone to make white, opaque glass. Threads of differently colored glass were often arranged in intricate patterns on the surface of the object being made, which was then, still hot and plastic, rolled gently over a flat surface, so welding the pattern of threads into it.

We saw earlier how silver was often smelted from lead ores, with the apparent wastage of the lead. Indeed, lead as a metal had been known from perhaps 3000 B.C., for it is comparatively easy to smelt, but little use had been found for it because it was far too soft for making tools and too unattractive for making personal ornaments. From about 1500 B.C. onward, however, we begin to find lead being used not as a metal but as an ingredient in both glasses and bronzes. Lead, if present in a sufficiently large quantity, alters the characteristic behavior of a glass on cooling. For example, glasses made only of an alkali, such as potash, and silica contract very considerably on cooling, so that if one attempts to use such a glass to cover the surface of, say, a pot or a brick, one finds that on cooling it contracts more than the material to which it has been

FIG. 136: *Egyptian glass vessel; about 1500 B.C.*

Glass now began to be shaped in the hot, plastic
state rather than being worked cold by abrasion.
Glass vessels were made either by dipping a sandy
clay core into crucibles of molten glass or by winding
threads of glass around a core, later reheating and
rolling the surface to make the threads fuse and to
flatten them. The surface of the vessels were often
decorated by applying threads of differently colored
glass which were, again, reheated and rolled into the
surface. The same decorative technique was used in
the making of beads and other small ornaments.

applied—as a result it cracks. But by the addition of large quantities of lead this shrinkage is greatly reduced, allowing one to apply
a glaze to an earthenware surface. This fact was evidently grasped
by the Mesopotamians shortly before 1000 B.C., when we find early
experiments in the glazing of brick and tiles, although for some unexplained reason little attempt seems to have been made to glaze
pottery at this time.

For the bronzefounder, the introduction of lead into his alloy
in small quantities allowed him to achieve what had hitherto only

FIG. 137: *Lead-glazed jug from Mesopotamia; about 800 B.C.*

been possible with a great deal of very careful forethought and preparation. The addition of between five and ten per cent of lead to bronze does not greatly alter its properties as a metal for the making of tools and weapons, but it does alter its properties when molten, for the alloy becomes far less viscous at low temperatures and thus easier to pour into an intricate mold. We know by examining earlier castings that ordinary tin bronze often cooled so rapidly that all the impressions of a mold were not taken up by the metal, leaving flaws in the casting that were either ignored or had to be filled later. By adding lead to the bronze this difficulty was largely overcome, and as a result not only did castings now become far more elaborate—indeed even weapons verged upon the baroque—but also new moldmaking techniques became prevalent.

To achieve really elaborate castings the object, perhaps either a small statue or the handle of a sword, was first modeled in wax. The wax was then covered with a layer of fine clay and after this a thick outer coating of coarse clays was added. The mold was allowed to dry and then heated so that the molten wax could be poured off while the coatings of clay became fired like pottery. The metal could then be poured into the hollow left after the removal of the wax so that the ultimate casting took on precisely the shape

FIG. 138: *The lead-glazed brick façade of a gateway,*
Babylon; about 800 B.C.

Lead as a metal alone was apparently too soft and
unattractive to have found many uses. The metal
was, however, employed as an ingredient of bronze,
glass and glazes. In the latter case, glazes began to be
applied, in Mesopotamia, to bricks and tiles, often
with additional coloring materials such as the salts of
copper. Although of a slightly later date, this glazed
brick façade gives a fair impression of the use to
which such bricks were put and of the effect they
could produce.

FIG. 139: *Diagram showing the stages by which Egyptian foundrymen cast a bronze statuette.*

FIG. 140: *Bronze statuette of a cat from Egypt; about 1500 B.C.*

With improved furnaces also came better methods of casting. Objects to be cast were often modeled in wax. The models were then covered with clay to form the mold, which was heated to remove the wax and fire the clay. The space left after the wax had been poured off was filled with molten metal and the mold broken open to recover the casting. By incorporating a core of clay within the wax model it was possible to produce a hollow casting, so as to economize on metal. The general principle of lost-wax casting is still used today in fine art foundrywork.

of the wax model, although, of course, the mold had to be broken open to get at the casting and thus could not be used again. The technique was also adapted to the making of small statues of bronze which, had they been of solid metal, would have been so extravagant as to be impossible. First a core of clay rather smaller than the intended statue was made and allowed to dry; to the surface of this was applied a layer of wax whose thickness would be the same as that of the metal in the finished statue and into which was modeled all the details. The core and wax were then coated with clay, dried and heated to remove the wax, and molten bronze was poured into the space left between the core and outer mold. Later the core might be removed, although this was not invariably the case. A life-size statue could have been made in this way, using eight or more pieces which were later joined. The result was far superior to the older technique of covering a wooden substructure with sheets of metal that had to be pinned in place. This technique of lost-wax casting in fact was not very different from that used by the bronzefounder today, although really large single castings remained beyond the early craftsman's ability.

THROUGHOUT THIS PERIOD of a thousand years many technologies remained virtually unaltered. Quarrying, masonry, the making of pottery and bricks and most of the techniques of agriculture saw little change. Apart from the technologies already described, most of the improvements that took place lay in the domestic field, in such things as weaving and the production of household luxuries.

It will be remembered that the earliest looms we know of were simple frames laid out horizontally at a short distance above the ground. In countries where it was practical to weave in the open air through most of the year this type of loom was perfectly satisfactory, but probably as the result of contact with more northerly people this period saw the introduction of a loom in which the frame was set vertically and thus could be housed indoors. The hanging warp threads had to be held under tension, and this was achieved by attaching a clay or stone weight to the lower end of the bundles of threads. Alternatively the warp threads may have been looped continuously around an upper and a lower horizontal beam so that the length of cloth that could be produced was in fact

FIG. 141: *Reconstructed drawing made from a damaged tomb painting showing a vertical loom, Egypt; about 1500 B.C.*

FIG. 142. *Vertical loom with weights holding the warp threads under tension as depicted on a Greek vase; about 500 B.C.*

just a little less than twice the distance between the beams. This arrangement was apparently unique to Egypt, for elsewhere in the Near East the warp-weighted loom now became the universal piece of weaving equipment.

Another device that we may almost certainly attribute to people who lived to the north of Mesopotamia was the plow with a proper share and sole. Hitherto the plow, even in Mesopotamia, where it was fitted with a small sole, had done little better than make a scratch in the surface of the soil. The plow with a share and

FIG. 143: *Simple vertical loom still in use today, from Jordan.*

The older form of loom in which the warp threads
were stretched at a short distance above the ground
appears to have been largely replaced in the Near
East by a loom in which the warp threads hung
vertically from an upper beam. In the Egyptian wall
painting the threads seem to have been held under
tension by a lower beam around which they were
looped. In all probability such looms varied little
from those still in use in northern Nigeria today.
Elsewhere, the warp threads were held under
tension by weights attached to their lower ends.
Although the weights made of stone or clay have
often been found, there are no contemporary pictures
of this type of loom, and to get an impression of what
it was like we have to look at later illustrations on
Greek vases. The vertical loom had the advantage of
occupying less ground space than the horizontal loom
and could thus be more easily housed indoors.

FIG. 144: *Drawing made from the impression of a Mesopotamian seal of about 1500 B.C., showing a plow with sole and single handle.*

FIG. 145: *Drawing of part of the decoration on a Greek vase of about 600 B.C., showing a plow with sole and single handle.*

flat sole was designed to dig far deeper into the soil and thus make a better furrow into which to plant seed. With the relatively light soils of Egypt and the Mesopotamian valley the older type of plow was perfectly satisfactory even if much of the seed was shallowly planted, but in the more northerly countries where the soils were heavier and the time required for germination of the seed was longer this type of plow could have been utterly useless. We assume that it was in these areas that the much heavier piece of equipment was devised. It appeared in Mesopotamia considerably before 1000 B.C. but in Egypt it was not adopted until nearly a thousand years later.

Even by 2500 B.C. we have some evidence that in Egypt oils were being pressed from various fruits. The machine used for this purpose was a simple bag press in which the fruit was placed in a cylindrical cloth bag, supported at each end by upright posts, which was twisted by two operators using simple bar handles.

FIG. 146: *Reconstruction of this type of plow based upon a number of decorated Greek vases and upon plows still used in Anatolia today.*

In Mesopotamia, but not in Egypt, the two-handled plow was replaced by one with a single handle and a far larger sole. This type of implement had the double advantage of making a rather wider and deeper furrow, while leaving the plowman a hand free with which to goad his ox team. He could plow alone without an assistant to do the goading. In subsequent centuries this was to become the most common form of plow in Greece and other Mediterranean countries, while in only slightly modified form it is still to be seen in use today.

Although presses of this sort were undoubtedly used for extracting the oil from olives, it is equally clear that they were being used to extract aromatic materials from other plants to be used as flavorings and as scents. Indeed, during this period there grew up a considerable industry and trade in cosmetics, for which purpose many flowers and herbs were deliberately cultivated. Merchant seamen from Egypt were certainly exploring parts of the east coast of Africa and the upper reaches of the Nile in search of exotic materials for this trade, while from the Mesopotamian delta it seems more than likely that merchantmen were occasionally traveling as far as India in search of spices.

These early cosmetics were very largely unguents made of animal or plant fats and oils, scented with a multitude of aromatic materials. Indeed the unguent bottle, either of stone, glass or clay, now became a familiar household object among the wealthier citizens of the Near East. But it was apparently in Cyprus that there developed an industry based upon the extract of the opium poppy, laudanum. In many parts of the Near East have been found small handmade bottles shaped like the dried head of an opium poppy. These jars bear a peculiar decoration which seems to imitate the scar on the face of the poppy which was made in order to extract the opium-bearing sap. To the wealthier members of society this drug was in all probability used much as we would use aspirin today: to relieve headaches and hangovers and even, perhaps, to keep the baby quiet. In any event it was sufficiently treasured to allow the bottles to be included in many graves as provision for the afterlife.

In the last few centuries before 1000 B.C. the whole of the Near East underwent a period of considerable turmoil due to the intrusion of people from Central Europe and northern Asia. Many writers have seen the spread of the knowledge of ironworking as a part of this breakdown of the established order in the Near East. While Greece, Anatolia, Persia, Mesopotamia and Syria were largely overrun by invaders, the Egyptians had to contend with an attack from land and sea by people who appear to have been dispossessed by the northern invaders. The Egyptians called these intruders, whom they managed to repulse, the People of the Sea, and they have left us one excellent record of a naval engagement that took place. This relief from the temple of Rameses III allows us, at last, to say something rather more concrete than we have been able so far to say about the design and function of shipping in the Eastern Mediterranean during this period. The Sea People are depicted as manning boats high in the bow and stern, the stem and stern posts incidentally surmounted by animals' heads. By contrast the Egyptians are seen using vessels in which the keel has been extended to form a ram, and in which the mast is now set amidships. In general appearance the Egyptian ships would seem to be very similar to those vessels already in use by Egypt's more northerly neighbors before 2000 B.C., and it is a matter of question as to whether the Egyptians either manned or built their own navy, whereas some authors have suggested that much of the Egyptian naval enterprise during these centuries depended upon mercenary sailors from the Levant.

Whether this was so or not, these tomb decorations show

FIG. 147: *Bottles in the form of dried poppy heads from Cyprus; about 1500 B.C.*

Juices were extracted from a wide variety of flowers, herbs and spices and compounded with fats and oils to provide unguents. Unguent bottles are a common feature of the homes and graves of the wealthy, although the precise nature of the great majority of these cosmetics and balms is quite unknown. The shape of one type of bottle, however, does betray its probable contents. Made in Cyprus, and closely copying the form of the dried head of an opium poppy, it is found widely throughout other countries of the Near East.

certain innovations. For example, the ships of both the Egyptians and their opponents are fitted with fighting tops to the masts, from which an Egyptian, in one instance, is clearly seen slinging a shot. Furthermore, the Egyptians were using grappling irons thrown from the ends of long ropes as a means of hauling ships closer together either to board them or perhaps as a means of bringing the ram to bear on the enemy's ship. The ships of both the antagonists are fitted with washboards above the gunwale and screens in the fore and after peaks to protect the crews from the seas.

The Egyptians, however, were perfectly familiar with Mediterranean merchant shipping, and from a tomb of a slightly earlier date than the temple of Rameses we have what is probably an Egyptianized view of these broad-beamed ships with a high bow and stern, vessels which appear to be only partly decked and on which may be seen some of the cargo contained in large jars, while above the gunwale runs a rather crude-looking railing.

FIG. 148: *Part of the decoration of an Egyptian tomb of about 1500 B.C., showing merchant ships in port.*

FIG. 149: *A modern model reconstruction of this type of ship. (left)*

Despite her very considerable overseas trade, Egypt adhered closely to ships whose design still echoed that of the reed boat. Although about seventy feet over-all, ships were still apparently without keels, and support was still given by a rope carried from stem to stern, as in the case of ships a millennium earlier (Figure 116). The cargo was carried on deck. A single mast was now stepped amidships, and the sails were furled by lowering the yardarm. The cargo in this instance includes elephant tusks, trees with their roots contained in baskets, and monkeys.

FIG. 150: *Part of an Egyptian tomb decoration of about 1200 B.C., showing the Egyptians repulsing invaders.*

It is difficult to determine from tomb paintings the size of a ship. Since as a convention the skipper is often depicted far larger than other members of the crew, the figures are not a very useful guide to scale. However, in recent years skindivers have discovered a wreck off the southern coast of Anatolia which goes some way toward helping us reconstruct this kind of vessel. In this case the ship had been carrying ingots of copper which apparently were laid directly into the hull of the ship, although separated from the timbers themselves by a substantial layer of brushwood. The ingots were fairly heavy, and one suspects that the brushwood was put there simply to protect the hull from damage should the cargo move. Although the ship has very largely disintegrated, the spread of the cargo on the sea bed suggests a vessel of some thirty or forty feet in length. For want of more reliable data, we must suppose that this would have been about the average-sized cargo vessel of the period, while this figure is to a degree supported by the fact that the ships shown in the temple of Rameses were rowed by seven oars on either side.

The ingots of copper recovered from this wreck are of a type hitherto occasionally discovered during normal archaeological excavation as well as being depicted in a few Egyptian murals, while their shape was used in Crete as a pictogram on records made on clay tablets. Each was a rather flat rectangle of metal

FIG. 151: *Part of the decoration on a vase from Greece, of about 1200 B.C., with missing areas restored.*

FIG. 152: *A reconstruction of how the ships of the Egyptian tomb decoration more probably looked.*

Of a rather later date is the picture of the Egyptian navy repulsing invaders from the Eastern Mediterranean. The Egyptian artist, probably more familiar with reed-built boats, appears to have distorted the shape of both types of ships by giving the keels an excessive curve. The Egyptian ships were more likely to have been long, low vessels with a ram, the rowers protected by washboards or awnings. They may even have been ships manned and designed by mercenaries, for they show a strong resemblance to vessels at that time in use in Crete and Greece. The invaders' ships seem to have differed only in detail from the earlier merchant ships from Crete (Figure 130). The Egyptians are seen using grappling irons, while both types of ships are fitted with fighting tops at the masthead—a feature seldom to be seen in Mediterranean shipping.

FIG. 153: *Copper ingots in the form of an ox hide from a wreck off the south coast of Turkey; about 1200 B.C.*

Copper was traded in the form of flat ingots of metal often provided with four protruding "handles," presumably to make handling simpler. A number of individual ingots of this type have been discovered, but those illustrated here came from a large group resulting from the underwater excavation of a wreck off the south coast of Turkey.

Ingots of this type are also often depicted in tomb paintings (see the man carrying an ingot, Figure 133, right) and even as an ideogram on Cretan clay records. In their wide geographic distribution and their comparatively uniform shape may be seen a step toward the idea of currency.

some two feet long by one foot wide with four protuberances at the corners. The shape is vaguely like that of the skin of an animal without head or tail; this has resulted in their being referred to as oxhide ingots, although the protuberances may have been added to the castings simply to make them more readily handled. Nevertheless, the ingots clearly represented real wealth and may have served as an early form of currency.

By the end of the period discussed in this chapter shipping had evidently become an important means of transport and communication in the Near East, although during the thousand years under review, due perhaps to a lack of information we cannot honestly say that there were many vast or great changes in ship design. The chariot, by contrast, is more often depicted and actual examples have been recovered from Egyptian tombs. During this period a number of changes took place in chariot design. Early chariots had an axle centered to the platform upon which the charioteer stood, so that he stood directly over the axle. The wheels were invariably four-spoked. Somewhere around the year 1300 B.C., however, two changes were made. First, the number of spokes was increased to six, which would suggest that a greater loading was being put on the wheels; and second, the axle was moved to the trailing edge of the platform. The charioteer was thus no longer balanced over the axle itself but his weight was carried partly by the wheels and partly by the yoke over the shoulders of the inner pair of horses. On the whole this seems a far more practical arrangement than that in the earlier form of chariot, for the weight of the charioteer could now be used to bear down upon the yoke and thus prevent its riding up to force the throatband against the windpipes of the inner pair of horses. The yoke, furthermore, was redesigned, for in its original form it was more suitable for the broad shoulders of an ox than for those of a horse. Into either end of the yoke was now mortised a new member in the shape of an inverted letter Y. The branches of this fork bore down on the shoulders of the horse, which were protected against rubbing by the addition of a small pad. Who was responsible for these changes it is hard to say, but yokes of the kind just described have been found in southern Russia and in China, while in the latter area many more than six spokes had been added to the chariot wheel well before 1000 B.C.

FIG. 154: *An Egyptian tomb painting of about 1200 B.C.,*
showing a chariot with wheels with six spokes.

FIG. 157: *Diagram showing how, by shifting the axle to*
the rear of the chariot, the weight of the crew
was carried to a greater degree by the yoke. (below rig▶

Toward the end of this period, in the twelfth
century B.C., changes were made in the design of
chariots. The wheels, which hitherto had only four
spokes, were given six spokes, while the axle was
moved from a position in the center of the cockpit
platform to its rear edge. Both changes were
seemingly designed to allow the vehicle to move over
rather rougher ground: the added spokes prevented
the wheel from buckling, and the change in position
of the axle meant that the weight of the crew was
partly carried by the yoke bearing down on the
horses' shoulders, thus correcting the seesawing
effect when moving over difficult terrain.

FIG. 155: *A Hittite relief of about 1200* B.C. *from southern Turkey.*

FIG. 156: *Reconstruction of this type of chariot.*

FIG. 158: *A relief from southern Turkey, of about 1000 B.C., showing a mounted warrior: one of the earliest representations of a horse being ridden.*

The supremacy of the chariot as a mobile weapon was soon to be challenged by cavalry. Enough had been learned about the control and training of horses to allow mounted men to fight from their backs. Early pictures suggest the use of a saddlecloth or even a rudimentary saddle, but stirrups seem not to have been thought of until another thousand years, when they began to be used by nomadic Asiatic people. Their more general adoption began with the invasions of nomads at the end of the period of Roman domination.

On balance, one suspects that the ideas were developed and diffused from the people of the Asiatic steppes.

With these improvements in harnessing and chariot design one might have expected the chariot to have become a rapid means of communication between one city and the next, but in fact neither in Mesopotamia nor in Egypt were roads taken at all seriously. Although within the cities the roads were often carefully leveled and paved, it would seem that outside the city walls the only roads to which any attention was given at all were those that led either to religious monuments or to quarries or other workings from which heavy burdens had to be carried. Good roadways had not even been developed for military purposes. In records of the period we often read of chariots getting bogged down and the fighting men forced to go on foot, also of baggage wagons being stuck in the mud. The same attitude prevailed regarding bridges. While a few were built in or near cities, elsewhere rivers had to be crossed at fords or by ferries, while for military purposes rafts were frequently used and less frequently pontoon bridges.

Thus, by 1000 B.C. throughout the larger part of the Near East the seas and the waterways still served as the principal means by which goods were transported over long distances. The wagon remained a slow, lumbering vehicle more suitable for shifting farm produce over short distances than as a means of overland transport, while the chariot despite its improvements could still be used effectively only in open country that was reasonably level.

However, from the same place of origin as the chariot there now arrived a new and more rapid means of getting about, for the people of the Asiatic steppe had learned how to train horses well enough to be able to ride them. It was in the face of bands of warrior horsemen that the people living around the coastal belt of the Eastern Mediterranean fled southward, only to be repulsed by the Egyptians. Other Near Eastern kingdoms were to suffer in the same way as the peoples of the sea. The kingdom of the Hittites in Anatolia was completely overrun, while another wave of people moved down into Greece and the islands of the Eastern Mediterranean, and yet a third, among whom were the Persians, overran the whole of the Iranian Plateau. A fourth group occupied a large part of the Levantine coast. Only Mesopotamia, at that time under the more or less unified control of the Assyrian rulers in the north and of Babylon in the south, held out against these intruders. Thus the beginning of the first millennium B.C. in the Near East was very largely a period which to the historian and to the archaeologist appears to be so chaotic that for two or three centuries the term "Dark Age" seems to be applicable. To those studying the history of technology, however, the period is one of very considerable interest: new ingredients had been added to the pot from which, once it had simmered down, totally new ideas were to emerge.

6/Of Greeks and Roads and Riding Horses

(1 0 0 0 – 3 0 0 B.C.)

I N THE PREVIOUS CHAPTER we saw how an understanding of the working of iron had developed in the mountainous area to the north of Mesopotamia and how, for reasons that are far from clear to us today, this knowledge had not been transmitted to the people of the Mesopotamian valley or indeed to any other people of the Eastern Mediterranean. However, with the breakup of the old established kingdoms and the scattering of so many peoples the knowledge of ironworking was now carried far and wide. Even so, the historian is faced with one of those problems of cause and effect which are so often impossible to answer. The people of the Near East may have taken to using iron for the manufacture of tools and weapons because they liked the material and because they found it cheaper to produce than bronze. On the other hand, with the disruption of trade resulting from the destruction of the older kingdoms, iron may have been used in the first place by many people as an ersatz material in lieu of bronze. Whatever the reason, however, the introduction of iron as a toolmaking material had the most enormous effect upon the artisan. Up to this time, for example, the stonemason had had to work with wooden and stone tools for the larger part, but by 500 B.C. there had been evolved a set of mason's tools, largely punches and chisels, which would not seem at all out of place in the hands of a modern sculptor. To a lesser degree the same is true of the woodworker. For chisels and saws are better made of iron, to which an edge is more easily put, than of bronze. Indeed, it would have been hard to find any craftsman working in the Near East who, by the year 500 B.C., had not to one degree or another taken to the use of iron tools; in many cases the metal was to completely revolutionize working methods.

During the first three or four centuries following 1000 B.C. the old traditions of technology were continued and enlarged to a great extent by Assyria, Babylon and Egypt, and one might truthfully say that during this time the newcomers into the Near East gradually assimilated these technologies from their neighbors. Both the Assyrians and Babylonians, however, spent much of their time fighting, either endeavoring to enlarge their empires or more simply defending what they already had. In the field of military technology there is little doubt that the Assyrians took the lead. In a number of early encounters the Assyrians learned a sharp and painful lesson from marauding horsemen, and so they very rapidly established their own cavalry. At the same time they appear to have concentrated on building heavier chariots than had been used hitherto; but this may be illusionary, for although their newer chariot wheels had a far greater diameter than those of earlier vehicles, nearly all the illustrations of Assyrian chariots would suggest that the larger wheels were made by adding a deeper felloe to wheels of the original dimensions. This change presumably allowed them to move over rather rougher ground, and we should also notice that at times the crew of each chariot was as many as four men.

The principal Assyrian military developments, however, appear to have been in the field of siegecraft and in machinery for breaking down city walls. Of these, the battering ram first made its appearance as a wagon covered with shields, with a fixed projection like a pig's snout. The whole vehicle evidently had to be rocked backward and forward against the city wall. Such a machine can hardly have been efficient even against walls of mud brick, and very soon the Assyrians learned to build an armored wagon in which the ram, a long beam, was raised and allowed to fall against the wall. The only defense against this vehicle of destruction appears to have been a kind of grappling iron which was lowered by the defenders of the city in an attempt to immobilize the ram. We have already seen how the Egyptians, some two thousand years before this period, had developed a siege ladder on

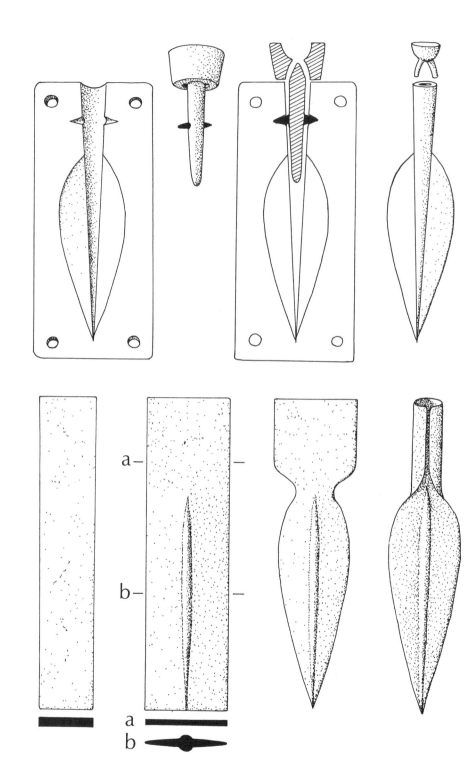

FIG. 159: *Diagram showing the very different stages followed in antiquity to manufacture a spearhead of bronze and one of iron.*

To a degree the use of iron changed design methods. Hitherto molds for casting bronze, or the patterns from which the molds were made, need not have been—and were sometimes clearly not—made by the foundrymen. By contrast iron objects had to be forged stage by stage, from a bar of metal, and their ultimate form thus fell within the province of the blacksmith. This distinction, and the effect it had on design, can be seen by examining the stages required in the manufacture of a bronze and an iron spearhead. In the latter case shaping was achieved largely by thinning down and folding the metal when red hot.

wheels. The Assyrians, now, probably following the success of their battering rams, began producing siege towers on wheels, often incorporated both ram and tower in the same vehicle.

The development of this rather crude military machinery was the direct outcome of the enormous success of the Assyrian army, comprising three main groups: infantry, chariotry, and cavalry. Within each of these classes was to be found a further division in which men were armed either with bows or with spears and swords, so that in any encounter, static or mobile, a screen of arrows could be put up allowing those armed with hand weapons to engage an already embarrassed enemy. Against such an onslaught few of the neighboring states could produce an army able to withstand an Assyrian attack in open battle. They had to take refuge either in cities or in the mountains. It was indeed in the eastern mountains of Persia that the Assyrians met with perpetual and incessant reverses, and it was from this quarter that military disaster was ultimately to overtake them.

It would be unfair to suggest that the Assyrians' only interest in technology was in the military field, for having once conquered a people, they developed the habit of removing the vanquished people's more useful citizens, particularly their craftsmen, to other centers, with the result that Assyrian cities literally teemed with

FIG. 160: *A Greek bowl, of about 400 B.C., decorated in low relief.*

FIG. 162: *Diagram showing various methods devised for lifting blocks of stone, (opposite) drawn from Greek sources; after 800 B.C.*

The introduction of iron tools made an enormous impact on many trades and industries, but especially on the working of stone. Iron wedges and hammers made the process of quarrying far simpler. Furthermore, because it was now possible to cut exact mortises in stone, it became possible to use a number of different devices by which to lift pieces of masonry, which in turn simplified the problems of building.

FIG. 161: *Drawing of a part of the scene from this bowl,
showing a blacksmith forging Hercules' club.*

The general use of iron for making tools and
weapons was accompanied by considerable changes
in metal working techniques. The fact that the metal
had to be worked when red-hot led to the
development of forges with built-in bellows fitted
with valves. Presumably the man standing by the
forge depicted in this scene is working the bellows,
the handle of which is in his left hand. To handle
and to shape the metal, tongs and hefty hammers
were needed as, too, was a substantial anvil.

FIG. 163*a*: *Part of an Assyrian relief of about 700 B.C., showing a heavy chariot.*

FIG. 163*b*: *Reconstruction of this type of chariot.*

The Assyrian response to the threat of attack by cavalry was to build larger and more rugged chariots. Wheels were given more spokes and enlarged by, apparently, the addition of a deep tire of wood. The crew was raised to four: driver, shield bearer, and two archers or men with hand arms. The number of draught horses was raised to four. Since the chariots could still be outmaneuvered by cavalry, they were often given the support of a number of mounted men.

FIG. 164: *Drawing from an Assyrian relief of about 800 B.C., and a diagram showing how the siege tower may have been operated.*

Unable to withstand Assyria's army, her enemies often took refuge behind city walls. The Assyrians, however, designed a number of siege engines which were not infrequently depicted on their reliefs. In its sophisticated form the siege engine appears to be a combination of a siege tower and a battering ram mounted on wheels. The battering ram appears to have been in the form of a pointed arm that could be used to prise out the mud brick from the city wall, while archers on the platform above protected those below demolishing the wall.

FIG. 165: *Part of an Assyrian bronze relief of about 900 B.C., showing a pontoon bridge.*

FIG. 166: *Part of an Assyrian relief of about 800 B.C., showing troops crossing a river on inflated skins.*

artisans drawn from other parts of their empire. This mixing of craftsmen from different regions should have resulted in cross-fertilization and in technological improvements; but, although they excelled in the making of glassware or glazed bricks, no really new technologies seem to have developed under the Assyrian regime. It is to the east and to the west of that country that we must now look.

By about 800 B.C. the situation in Greece and the Aegean islands had settled down. The newcomers, the Hellenes, initially an agricultural people, had learned much from the original population, whom they had clearly not totally eliminated. From our point of view the most important single technique that they mastered was how to build and to sail ships. Their vessels appear to have differed very little from those of the people they had conquered. Although they chose to depict them in a rather different manner, the Greeks are to be seen either rowing the identical low craft with a ram bow and a high, upturned stern or sailing the same type of craft with high stem and stern as had their predecessors in this area. These vessels were to result in the Greeks' not only becoming merchantmen but also, later, as their population expanded, great colonizers. This in turn brought them into contact with all the peoples not only of the Eastern but also of the Western Mediterranean.

The shipping trade in the Mediterranean, however, was not the sole monopoly of the Greeks, for they shared it with the people of the Levant, particularly the Phoenicians, who themselves were equally great colonizers, Carthage on the North African coast being the most successful and best known of their settlements. But the Phoenicians differed from the Greeks in one important respect: they were Asiatics and their trade was carried out with and for Asia—with Assyria and other people in the eastern end of the Mediterranean. To the Phoenicians, presumably, Assyria and its neighboring countries were the center of the world, and for them, except in matters of shipping, it was unthinkable that anybody elsewhere could produce manufactured wares or goods better than they. By contrast, the Greeks were evidently not quite so convinced of Assyrian technological supremacy, and by the end of the

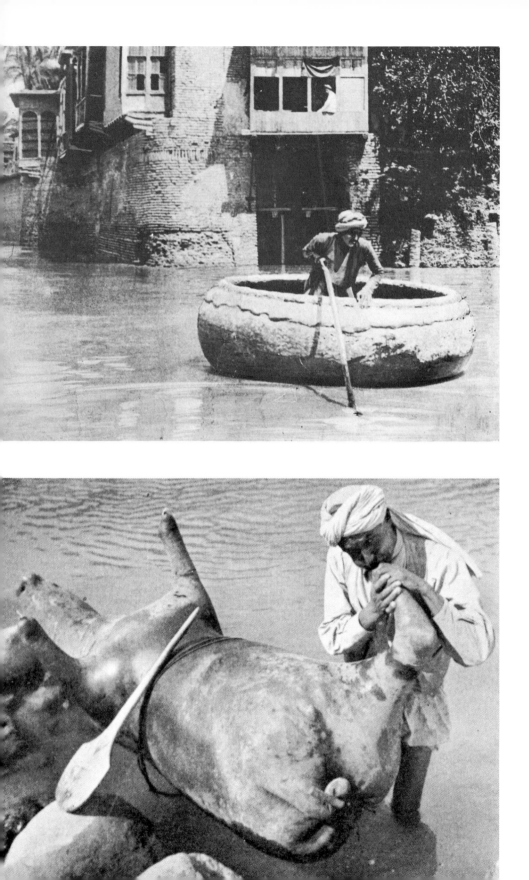

FIG. 167: *A gufa—the biblical coracle—in use on the Tigris today.*

FIG. 168: *Inflating a goatskin float, Nepal. (below)*

> For crossing rivers the Assyrian army resorted to
> a number of stratagems. At times inflated skins were
> used as floats, and equipment, such as a chariot, was
> taken to pieces and carried across the river in sections.
> Elsewhere a pontoon bridge would be constructed
> using as floats large basketwork boats made
> watertight with bitumen. Curiously, both inflated
> skins and the simple basketwork boats—*gufas*—
> are still in use in the Near East today.

seventh century B.C. the Greeks were already demonstrating their potential superiority in many fields.

The most obvious example of the advances made by the Greeks in the technological field is that of their pottery. The superiority of Corinthian and Attic wares to anything being produced elsewhere in the Western World at this period is without question. But this pottery did not just happen. Behind it lay a number of technological innovations that the Greeks clearly devised themselves. To begin with, the potter's wheel was no longer a small turntable placed close to the ground but had become a large flywheel with a working head raised above it to a height of a foot or eighteen inches above the ground. Sitting on a stool, the potter shaped his vessels on the wheelhead while an assistant, probably an apprentice, sat opposite, turning the heavy flywheel by hand. This heavier and therefore steadier wheel, we shall see, was to be of considerable importance later. Second, once the pottery had been shaped and dried it was replaced on the wheel—probably inverted—and shaved, thus producing a much finer surface than would have been possible simply by throwing it on the wheelhead. This habit of shaving pottery does not appear to have been carried out elsewhere at this time, and this practice was probably to have an effect in a completely different field of technology. Finally, the contrasting colors of black and red were produced by a very sophisticated process demanding the production of an extremely fine clay material, followed by a fairly elaborate sequence of firing. The clay slip used for producing the black areas of this pottery was

FIG. 169: *Reconstruction of the Greek "longboat" of about 600 B.C., made from a drawing on a vase in the Louvre.*

This drawing of a Greek "longboat" is probably about as near as one can hope to approach to a blueprint. The picture is part of a scene on the "François" vase in the Louvre and is unfortunately not complete, a small section being missing just abaft the bows. When reconstructed the vessel appears to have been between thirty-five and forty feet long with a freeboard of about eighteen inches, a ram bow, a high stern, and an awning in the forepeak. The hurdlework above the gunwale suggests a crew of twenty-six oarsmen. The original drawing makes it perfectly clear that the oars were placed immediately over the gunwale. If, on the other hand, a spare crew were seated inboard of the oarsmen, each man provided with an additional longer oar, both crews could row simultaneously in times of stress, the inboard crew rowing over the hurdlework. This would have brought the total crew up to fifty. In fact this vessel is probably the penteconter, the fifty-oared longboat, known otherwise only from literary sources. The original drawing was made shortly after 600 B.C., but earlier less reliable drawings suggest that for at least two centuries before vessels of this general character had been in use among the Greeks.

made by stirring clay with water and an alkali, probably leach from wood ash, and allowing the material to stand, when the very fine clay particles alone would remain suspended. This suspension was then decanted and allowed to evaporate until it achieved the required consistency with which to paint on the decoration. The wares were fired in a domed kiln to a temperature little short of 1000° C; at this point the openings of the kiln were closed down, the momentary effect being to blacken the entire surface of the

FIG. 170: *Limestone figure of a potter from Cyprus; about 600 B.C.*

FIG. 171: *Reconstruction of the Greek potter's wheel of about 600 B.C., made from a number of painted vessels of this period.*

Even as late as the sixth century B.C. Greek potters depicted themselves seated on a low chair, shaping their wares on a large turntable with a raised central disk on which the clay was worked. Either the potters turned the wheels by hand themselves, or a youth sat on a stool opposite to perform this duty. So far the wheel showed very little change from that used much earlier in Egypt. Unfortunately, shortly after this date potters no longer depicted themselves on their wares, with the result that we do not know precisely when the kickwheel was introduced.

FIG. 172: *Reconstruction of a Greek kiln of about 500 B.C., made from a number of decorated plaques from Corinth.*

FIG. 173: *Fragments of a decorated plaque from Corinth with the remainder of the scene reconstructed.*

At first sight Greek potters' kilns seem to show little change from those in use in Mesopotamia two thousand years before (Figure 51). On the other hand, the fire opening had been extended, presumably to allow a larger space for combustion and to improve draught. A loading door was usually shown with a spy hole through which the potter could see what was going on inside the kiln. Frequently the potter depicted himself climbing onto the dome of the kiln to open or to close the top vent, or raking out the fire—vital steps in the production of black-and-red wares.

wares. When the kiln had cooled to a little below 800° C. the apertures were reopened, so readmitting air, with the result that those areas which had been covered with the fine slip remained black while those left untreated gradually lost their blackness and became a clear red. Stated in this bald way the process seems simple, but to achieve the end result a very great deal of observation, trial and error were necessary. Indeed it is only in recent years that the method used has been re-established by scientists eager to determine how it was done.

Corinthian and Attic wares were traded widely throughout the Mediterranean and ultimately found their way even into Central Europe. Although we have no idea who developed this interesting system of potterymaking and decoration we are more fortunate in other fields, for we come now to a period in which historical records begin to be more reliable. In the small Greek states on the western coast of Anatolia, particularly in Ionia, is recorded a group of people who appear to have devoted themselves wholeheartedly to technological improvements. Of these, Thales of Miletus is probably, undeservedly, the best known. In his travels Thales had evidently seen Egyptian surveyors at work and had studied their methods of land measurement and the equipment they used. It was also probably from this source that he learned of astronomy and how the Egyptians used the stars to determine positions. Armed with this knowledge and a great deal of practical common sense, Thales devoted himself to the study of navigation at sea. By a system of triangulation, in which, of course, the magnetic compass was not used, he evolved a method of determining the distance of ships at sea from dry land, and from his knowledge of astronomy he laid at least the foundations of navigation by the stars.

A contemporary of Thales, Anaximander, was evidently equally interested in navigation, for he allegedly produced the first map of the world. Maps of one sort or another had existed long before Anaximander's day. These were often crude estate maps or merely sketches to show how to get to such and such a point. Although they often stated distances and areas, they were seldom drawn to scale. The known world in Anaximander's day

FIG. 174:
*Reconstruction of a gromar from fragments found
at Pompeii.*

Simple surveying equipment such as this, which
allowed the setting-out of accurate right angles, had
long been in use in Egypt. It was, however, the
Ionian Greeks who first appear to have used
surveying equipment for the purpose of navigation.

was admittedly not very large, but nevertheless this was for one
man a gigantic undertaking and as a concept an enormous advance. Like Thales, Anaximander was also interested in the behavior of the heavenly bodies and it is to him, too, that is given
the credit of inventing the first sundial, more probably a simple
astrolabe. Once again, Anaximander was clearly drawing on knowledge that already existed, for sundials of a fairly simple character
were in use elsewhere before his day, but he attempted to give
the device some sort of correction for the varying seasons, so providing a practical timepiece.

Two other inventors of this period—Anarcharis the Scythian

FIG. 175: *An Egyptian relief (left) of about 300 B.C.,
showing the god Khum seated at a kickwheel.*

FIG. 176: *Reconstruction of this type of kickwheel.*

and Theodorus of Samos—have been less kindly treated by historians than Thales. Between them they were once credited with improving the anchor, the bellows, and the potter's wheel; of inventing the lathe and the key; and of bettering a number of mathematical instruments as well as devising a method of casting bronze. This is an impressive list that would have done credit to, and made the reputation of, any inventor. The difficulty in accepting the statements lies in the fact that many of the objects and methods in this list were known to have existed before Anarcharis and Theodorus, and it has been assumed these men merely filched ideas from Egypt, Assyria and elsewhere and brought them back to Greece. To some extent this may have been true, yet the evidence suggests that the improvements listed were made during this period. We have just seen that the Corinthian potter worked on a wheel which was fairly close to the ground, requiring an assistant to make it turn continuously. Following this we have no other evidence of what potters' wheels looked like until the

FIG. 177: *Pottery figure made by John Broad in 1883.*

By 300 B.C. the working head of the potter's wheel had been raised as, too, had the height of the potter's seat. He could now sit at the wheel and kick it with one foot to make it rotate, so leaving both hands free to shape his wares. These changes in design were possibly the work of Anarcharis, who is alleged to have "invented" the potter's wheel. With very little modification this type of wheel was still in use in Europe as recently as the end of the last century.

Ptolemaic period in Egypt some three hundred years later, when we find the potter seated on a high bench, the wheelhead raised to waist height and the potter himself rotating the wheel by kicking it with his right foot. It seems therefore quite possible that Anarcharis saw the Corinthian potter's wheel, realized the defects in the device and suggested certain improvements. By raising the wheelhead and the height of the seat the potter was able to make the wheel rotate by kicking it himself. Anarcharis is also said to have improved the anchor. Certainly up to this point the only anchors known were large stones drilled at one end to take the mooring rope and at the other end to take short wooden flukes. Such anchors would have dragged in heavy seas. After this period we find in use the conventional anchor with developed flukes; once again, it seems probable that Anarcharis saw that an anchor could be prevented from dragging by redesigning it in the form of a grapple, with which he may well have been familiar. His third invention, the bellows, already existed as hand- or foot-operated skin bags, or as skin-covered drums. After this period, however, we find that the bellows with a clapper valve was in use, and once again there is no reason to believe that Anarcharis was not responsible for the innovation.

The earliest lathe that we know of comes, again, from a Ptolemaic tomb painting, and thus when we read of Theodorus of Samos having invented the lathe there seems no good reason to doubt this statement, especially when we recall that he may have seen Corinthian and Attic potters shaving their wares on the wheelhead, a process which could easily have suggested to him how a piece of wood might be shaved down in the same way, were it made to rotate. It is more difficult to say what is meant by the statement that Theodorus invented a method of casting bronze. But yet another Ionian inventor, Glaukos of Chios, is reputed to have invented the welding of iron, that is to say, the joining of one piece of iron to another by heating it to a good red heat and hammering it on an anvil. Certainly until this time blacksmiths had avoided this method and had often gone to peculiar extremes in order to join iron by means of such devices as rivets and lapped flanges. With the development of the bellows, allowing a good red heat to

FIG. 178:
*A lathe as depicted
in an Egyptian low relief;
about 300 B.C.*

FIG. 179:
*Reconstruction
of this type of lathe.*

FIG. 180: *A similar simple lathe (opposite) as used by the
Hunza today.*

The lathe, the invention of which was again
attributed to Anarcharis, is first depicted in the
third century B.C. in Egypt. The object being turned
and the spindle were held horizontally between stocks.
A cord was passed round the spindle, and one
operator made the spindle and workpiece rotate back and
forth by pulling on the cord. A second artisan cut
back the workpiece with a chisel. The idea of the
lathe may well have been derived from potters
shaving down their dried wares on the wheel.

FIG. 181: *Diagram showing how the lathe was adapted for stone and glass cutting by the addition of a lap wheel.*

FIG. 182: *A Greek vase about 600 B.C., showing an anchor being thrown overboard.*

FIG. 183: *Another form of stone anchor in use from about 1000 B.C., to which wooden flukes were probably fitted.*

FIG. 184: *Anchors as depicted on lead weights from Syria; about 200 B.C.*

In the seventh century B.C. many ships' anchors were still little more than large stones drilled through the middle to take the hawser. Anchors of this kind must have dragged in heavy seas. To overcome this defect an anchor in the form of a triangular stone was developed. The hawser was tied through a hole in the upper angle of the stone, while a pair of wooden flukes were fitted through holes near the lower edge. Traditionally the anchor, as we understand it, was invented by Anarcharis, and certainly by the second century B.C. the anchor had attained the form in which we know it today.

be achieved easily—coupled with the development of really heavy anvil tools (as depicted on Attic and Corinthian vases)—there is no reason to doubt that Glaukos may well have improved the technique of welding. Indeed, the evidence suggests that the Ionian Greeks were giving all the processes of metalworking a critical eye, the results of which are very evident. Their bronze statuary, for example, apart from any aesthetic qualities, shows a considerable technological advance over anything being produced elsewhere in the Near East at this time, for it is made of fewer and larger castings than hitherto and incorporates such devices as iron armatures for its support.

However, what is more interesting about the Greeks of this period is not whether certain inventors were responsible for what has been claimed for them, but the mere fact that their names are recorded as having made those inventions. Plainly the attitude of the Greeks to technological innovation as reflected in these statements is very different from that of the Asiatics, among whom a craftsman remained for all his skill and all his inventiveness a mere craftsman. That this attitude was short-lived we shall soon see, but for a few centuries at least the status of the technician among the Greeks was raised far above that of his fellow technicians in Asiatic countries. To be interested in technology was respectable, and an inventor was regarded as a benefactor. The Greeks, however, did more than metaphorically raise the level of technology. They raised it physically as well.

A study of pictures of craftsmen in early Egypt and Mesopotamia will show them either standing at their work or kneeling or squatting on the ground: essentially the ground was their workbench. By contrast, when the Greek artisan is not shown standing, more often than not he is seen seated on a stool or a chair, his work being carried out on a low bench or table. At first sight this may seem insignificant, and one could argue that environmental differences—the kinder weather in countries other than Greece—would account for the choice of working position. This may well be true, but it does not alter the ultimate fact that many crafts are far better carried out on a bench than at ground level, such as woodworking, masonry and fine metalworking, to mention but a

FIG. 185: *Egyptian sandalmakers, as depicted on the walls of a tomb; about 1500* B.C.

FIG. 186: *A Greek sandalmaker of about 600* B.C., *from a vase.*

FIG. 187: *A shoemaker in Jordan today.*

Unlike his Egyptian and Asiatic counterparts the
Greek artisan does not seem to have enjoyed working
while squatting on the ground. More often than not
he is shown seated on a chair and working at a
table or bench. As has just been seen in the case of
the potter's wheel, this difference in posture allowed
the development of tools and techniques that would
have been impossible had the working position
been one of kneeling or squatting. Throughout a
large part of Asia and Africa today craftsmen still
show a preference for working close to the ground.

FIG. 188: *Part of an olive press from a Greek relief of the sixth century* B.C.

FIG. 189: *Reconstruction of this type of beam press, based on this relief and a number of painted vases.*

FIG. 190: *Diagram showing how the pulley and the screw were later used to work the beam of this kind of press.*

Olive oil was to become one of the main Greek exports. Initially the oil was extracted in a simple beam press, of which a few illustrations are known. Thales, who was in reality the father of navigation, is also remembered as once having "cornered" the market in olive oil.

few. Indeed it seems unlikely that many of the iron tools evolved during this period for particular crafts could have been developed at all had the craftsman continued to work at ground level.

During the period immediately preceding the seventh century B.C. the Greeks had, agriculturally speaking, slowly adapted themselves to their new environment and had developed the cultivation of the vine and olive to such a degree that olive oil and wine, together with their pottery, became their principal exports. There can be little doubt that it was this enormous volume of trade that virtually forced the people of the Mediterranean into adopting a medium of exchange, rather than continuing to base their commerce on direct barter. Reputedly, Croesus of Lydia, a small state in northern Anatolia, introduced a form of currency in about the year 700 B.C. This ruler introduced the first true coins, pieces of metal stamped with a guaranteed weight and purity. A hundred years later the Greek cities were all issuing their own bronze and silver currency, a practice that was soon to spread throughout the civilized world.

One cannot, of course, issue currency of guaranteed purity unless one has metallurgists capable of producing pure metals. Nor can the coins be of a guaranteed weight unless one has

(a)   (b)   (c)

FIG. 191: *Coins (staters) from: (a) Ionia, of about 600 B.C.; (b) Aegina, of about 575 B.C.; (c) Thasos, of about 525 B.C.– Example (a) is made of electrum, (b) and (c) of silver.*

With the growth of trade in such commodities as pottery and olive oil, the demand grew for an acceptable token of exchange. Coins, pieces of metal of guaranteed weight and purity, stamped with the mark of the issuing authority, first appeared in the seventh century B.C. Early coins were fairly simple in design, but by the sixth century most countries of the Eastern Mediterranean were minting their own coins, the designs of which showed increasing complexity.

weighing devices capable of accuracy. These last two observations may seem rather trite, but they do show, nevertheless, how far the Greeks had advanced by this period. To be able to produce pure silver coinage they must have been familiar with methods of cupellation, in which the impure silver was heated either in a furnace or in a crucible with materials capable of absorbing such impurities as might be present, principally lead. From a slightly later period we have a number of descriptions of this process which show that by then it was old and well established. Furthermore, by cupellation with common salt, silver could be extracted from electrum, the naturally occurring gold-silver alloy, so allowing the production of reasonably pure gold coins.

The development of a monetary system speaks for itself. The volume of trade being carried out by the Greeks and other countries of the Near East was obviously on the increase, and so far as the Greeks and the people of the Levant were concerned, this trade was being conducted by sea. We might reasonably expect, therefore, to find at this period that the Greek shipping and that of the Phoenicians were fairly identical. But this is far from the case, and the very differences themselves say a great deal about the attitudes of these two different peoples. By about 700 B.C. the people of the Levant coast had made one very important change in the placing of the oarsmen in those ships that were rowed: hitherto there had been a single line of rowers, the oars being placed over the gunwales while the rowers were protected by washboards, as we have already seen. In time the washboards had become incorporated in the hull of the ship, which was then built with a row of portholes through which the oars were placed. More simply, the ship's sides were extended upward and rowing took place through ports. But the disadvantage of this system was that the number of oarsmen was restricted by the length of the vessel. The Phoenicians, however, found that they could increase the number of oarsmen by placing a second row of seamen slightly above and inboard of those rowing through the ports, their oars

FIG. 192: *A stone relief from southern Turkey, of about 800 B.C., showing a warship during battle. It is manned by a single bank of oarsmen.*

FIG. 193: *An Assyrian relief of about 700 B.C., showing a ship of the same general form, but manned by two banks of oarsmen.*

FIG. 194: *Conjectural reconstruction of an Assyrian bireme of about 700 B.C., based on this and similar reliefs.*

FIG. 195: *A Greek bireme as depicted on a vase; about 500 B.C.*

During much the same period that the Greeks were developing the penteconter, the people of the Levant made a number of changes in the design of their ships. The washboards, in use well before 1000 B.C., were incorporated into the ship's side so that the oarsmen were in effect rowing through a line of ports. A second bank of oarsmen were then able to row over the gunwhale. This two-bank arrangement, the bireme, was a permanent one unlike that suggested for the penteconter, which was presumably inspired by the Levant biremes. The hull appears to have been shorter than that of the Greek vessels; a superstructure was added to it to give advantage to the armed men. This gives the vessels the appearance of having been unduly top-heavy.

FIG. 196: *Diagram showing the rigging used on Greek and Levantine shipping of this period, based on reliefs and painted vases.*

By 500 B.C. the Greeks, too, were building biremes with one bank of oarsmen rowing through ports and the second over the gunwale as in the Levantine manner. Illustrations of rigging of this period show that the Greeks and Phoenicians were using identical systems. Unlike earlier Egyptian sails, furling was now done by hauling on sheet lines which were secured either to the foot of the mast or to the gunwale. The yardarm could be lowered by means of a pair of hoists that ran through a block at the masthead.

protruding over the gunwales of the extended ship's side. This system virtually doubled the number of oarsmen without extending the length of the ship, and one rather suspects that this scheme was the outcome of a policy which demanded that the keel should be a single long timber. There are a sufficiently large number of pictures of galleys of this type to suggest that the number of oars employed on each side, even in double banks, never exceeded twenty, and that the vessels were frantically top-heavy, especially when they were used as men-of-war, at which times a deck had to be provided at a raised height to enable the archers and other armed men to fire upon their enemies. Furthermore, it seems unlikely that such ships, with their severely limited number of oarsmen and their great top weight, were able to operate in heavy weather. By contrast, Greek vessels of the same period seem to have become excessively long and rather low in the water, frequently being depicted with twelve or thirteen men rowing on each side, the oars projecting over the low gunwales. Above the gunwale is often shown a light framework of wood; in early paintings particularly, this is distinctly shown as being lashed together. This framework would be quite useless as a protection against the weather and, in any case, is never shown covered with an awning, although other parts of the ship where awnings are used are quite clearly illustrated. We know from literary sources that it was at this time that the Greeks introduced a vessel they called the penteconter, the fifty-oared vessel. It seems likely, therefore, that the ships that we see depicted with twelve or thirteen oars on each side were normally rowed by only half the crew while the other half were resting. In times of stress, however, when great speed was required for fairly short periods, the resting crew, seated as in the double-banked Phoenician ships, could put out rather longer oars over the hurdle work, doubling the number brought into play and thus raising the total number of rowers to fifty. Unlike the Phoenician vessels, such a ship would have been capable of countering the strong currents of the Hellespont and so entering the Black Sea, and it is important to notice that it is from this period onward that Greek trade with the Black Sea coasts began to take on a major role in the country's economy.

At this time the Phoenician ships were rowed by slaves, whose cramped conditions mattered little to their masters. By contrast, Greek shipping was manned by freemen and the conditions under which they worked were probably the subject of debate—a matter at which the Greeks excelled. Significantly, when the first pictures of double-banked ships of Greek design first appear, around 500 B.C., slavery had already become a major aspect of Greek economy, although the oarsmen were still, technically, freemen.

We must now turn our attention to the progress of affairs on the mainland of the Near East. In 625 B.C. the Babylonians defeated their Assyrians neighbors to the north. It may seem strange that the Assyrians with their magnificent war machine were defeated by their apparently weaker neighbors, but in fact the Babylonians cannot take all the credit for the defeat. They were helped in no small degree by newcomers into the district, an alliance of tribesmen, lightly armed, but superb horsemen, who formed a core of light cavalry with which the Assyrians found themselves unable to cope. Among these people were the Medes, who were to figure later in the history of this part of the world. For the moment this change in Mesopotamian rulers made very little difference. Babylon continued the traditions of Assyria, and among the people they moved into captivity—many of whom were craftsmen—were the Jews. The new Babylonian empire, however, was not to last long, little over a century, in fact, for in 538 B.C. the

Babylonian rulers in turn were overthrown. A second group of tribesmen, equally expert horsemen, the Persians, in alliance with the Medes, overran the whole of Mesopotamia, having first dominated the Iranian Plateau. Hence the whole of Mesopotamia, Persia and a large part of what is today Turkey fell into the hands of a single group of rulers.

It is very easy to underestimate the contribution the Persians made to technological advance. They had become in their outlook typically Asiatic, and politically they ran their empire in much the same way as had the Assyrians and Babylonians before them. The Persians held, however, a vast area stretching ultimately from the Punjab to the Mediterranean, and possibly because of this made one very great contribution to mankind. They developed a system of communications based on roads maintained by the central authority. These roads had posting stations at regular intervals, and by virtue of this messages carried by relays of mounted horsemen could be transmitted over great distances in remarkably short time. Many of these roads were paved, although, of course, in mountainous areas there were long stretches that were not. They all shared one thing in common, however: they had to be maintained in a state that would allow horsemen to move along them rapidly. While by today's standards we might not think very highly of these roadways, they were, nevertheless, a vast improvement upon the haphazard trackways and byways with which previous rulers of these areas had been content.

By 500 B.C. the stage was already set for a major conflict. On the one hand were the Persians ruling a vast Asiatic empire centrally controlled and capable of mustering enormous technical resources. On the other hand were the Greeks, a young, vigorous seafaring people, technologically on the move, anxious to increase their trade and, because of their rising population, create more and more colonies. It was only natural that the Greeks who had already settled on the western coast of Anatolia and the islands close to it should revolt against a Persian overlordship. Much of the first quarter of the fifth century B.C. was, therefore, devoted to resolving the rivalry of these two great peoples. It ended essentially in a stalemate. The Greeks beat off a threat of invasion from Persia.

The Persians found themselves unable to curb Greek commercial enterprise.

If this war can be said to have achieved anything at all, it acted as a stimulus to the Greeks to build better shipping. As early as 600 B.C. other seagoing people of the Eastern Mediterranean, more notably the people of Syria and Egypt, had begun to build large ships with three banks of oars, the so-called triremes; the Greeks, threatened with the possibility of an invasion by the Persians, for whom these other people of the Eastern Mediterranean acted as mercenary seamen, began also to build triremes. Much has been written about the trireme, but the truth is that we know very little about its construction or even how it was rowed. The sole evidence upon which we must base our reconstruction of them lies in a few literary sources, some very fragmentary marble statuary and a single rather crude clay model, a few paintings and some sketches which are possibly, but not certainly, intended to depict triremes. What does seem certain, however, is that the total number of rowers varied from 120 to rather more than 200 and that they were arranged in three banks, of which those in the middle bank sat farther inboard than the others. Finally, the literary evidence shows us that they were far from stable and that when ramming, which was the principal means of attack, in order to steady the ship the blades of the oars had to be allowed to drag in the water. We frequently read, also, of ships being swamped in heavy seas and of rammed ships capsizing. Nevertheless, there can be no question of the length of the trireme being restricted by the use of a single timber for the keel plate: the keel must have been composite. Although we have no knowledge of construction methods, the shipwrights were obviously now capable of building not only longer men-of-war but also mercantile shipping capable of carrying large cargoes.

Larger and better shipping meant, of course, that the Greeks were able to defeat the Persians at sea and remain a major sea power in the Eastern Mediterranean, although in the Western Mediterranean they were to become increasingly in conflict with Carthage and later still with Rome. In the meantime the Greek and Persian wars brought the Greeks into far closer contact with the people of the Near East than they had ever been before. From

FIG. 197: *Drawing made in the nineteenth century of a relief (since destroyed) from the Acropolis, showing a section of a trireme of about 500 B.C.*

FIG. 198: *Modern model reconstruction of a Greek trireme based on reliefs and literary sources.*

FIG. 199: *Clay model of a trireme from Egypt; probably about 500 B.C.*

FIG. 200: *Reconstruction of this type of trireme.*

Triremes, ships with three banks of oars, were being built in Egypt and the Levant by 600 B.C. A century later the Greeks were building them, too. We know very little about their construction. Literary references give the over-all dimensions and the number of oarsmen, while marble reliefs, such as that from the Acropolis in Athens, show the spacing of the oarsmen. Students of early shipping have suggested that the upper bank of oarsmen seen in the drawing of the Acropolis relief were rowing through an outrigger; the model in the Science Museum has been made to incorporate this feature. The clay model from Egypt suggests that in countries other than Greece triremes were not fitted with outriggers. The reconstruction shown here assumes that the Levantine trireme was little more than a bireme (see Figure 193) with the addition of a third bank of oars.

them they learned little that did them any good, for their newly acquired wealth added impetus to a social change that had already begun to take place. Even during the Persian wars Greek shipping was manned by freemen, but this was not a state of affairs that was to last for long, and soon we read of slaves being used to row Greek shipping. From then on the social structure of Greece became more and more similar to that of the Orient. Very soon all industrial enterprise was based upon slavery, a state of affairs that Greek philosophers bent over backward attempting to justify.

FIG. 201: *Reconstruction of a Greek merchant ship of about 500 B.C., based on a number of painted vases of the period.*

The penteconter, bireme and trireme were in essence men-of-war, although they may all have been used for trading in times of peace. The Greek merchant ship of 500 B.C. was, however, a very different vessel. It was clearly developed from the Cretan type of ship already discussed (Figure 130), and was probably only fractionally larger. It is often shown with a fiddle bow. One finds few illustrations in which the cargo was clearly stowed on deck and must assume, therefore, that the hull was more sturdily built than in the earlier Cretan ships, also that the cargo was carried largely in a hold.

7/The Late Greek and Roman Engineers and Their Devotion to Machinery

(300 B.C. – A.D. 500)

THE YEAR 300 B.C. represents in antiquity almost the end of one aspect of technological development in the Near East and indeed in the whole of the Western World. It is true to say that virtually no new raw material was to be exploited for the next thousand years and that no really novel method of production was to be introduced. What new advances were made were to be almost entirely in the field of engineering, and most of the principles involved had themselves already been discovered and applied, although usually on a smaller scale. Thus, during the entire period in which Rome dominated the civilized Western World one of the few really novel means of production that appears to have been developed in any industry was that of blowing glass. Hitherto glass vessels had been made either by dipping a core into molten glass—the core later being removed—or by winding rods of heated softened glass around the core and finally rolling this shape backward and forward over a flat surface to smooth it. The process of glass blowing had, in fact, to wait for the development of iron technology, for of the metals known in antiquity only an iron blowtube would serve for the purpose of blowing glass. It was not until the beginning of our era that glass blowing, therefore, became a widespread practice, but while the Romans may well have been responsible for the spread of this technique certainly no Roman ever invented it. It seems to have developed either in present-day Syria, or in one of the countries adjacent to it.

FIG. 202: *Diagram showing the stages by which a Palestinian "wineglass" was made in the first century B.C.*

Historians reviewing this state of affairs, this falling off in the rate of technological innovation, have tended to attribute it to a number of causes, the first of which was the widespread use of slave labor. The dirty end of production, it is argued, was put entirely into the hands of slaves and increased output could be achieved only by one of two means: either by acquiring more slaves or by making those one had work harder. Since it is not in the nature of slaves either to invent new means of production or to exploit new materials, the possibility of any further technological development came to an abrupt end. Alternatively, it has been argued that in the early centuries of the empire, that is to say until the beginning of the decline of Roman power, there was an excess of manpower and that any improvement in production would

FIG. 203: *A "free-blown" glass bottle of the first century* A.D.

FIG. 204: *A "mold-blown" glass bottle of the first century* A.D.

The blowing of glass vessels, which was introduced in the second or first century B.C. in the Levant, was to replace almost completely the earlier method of molding glassware. Either vessels were free-blown— the diagram shows the stages by which a plastic bubble of glass was converted into a "wineglass"— or the glass was blown into a piece mold as, indeed, is most of the glassware made today.

merely have resulted in massive unemployment, with which the rulers of the various provinces would have been unable to cope. Technological development was therefore actively discouraged.

In reading what was written by philosophers and statesmen of this period there is much in their writing to support this point of view. Indeed, the gist of what passed for philosophy, insofar as industry was concerned, was that technology was a filthy business fit only for slaves and that no intelligent thinking man would bother his head with it. But, for reasons that none of the philosophers were able to make quite clear, it was apparently quite in order for a member of the upper crust to concern himself with *engineering*, in its widest possible sense.

A third contributory cause that historians have often ignored is the appearance of a very mediocre and often dishonest class of lower-grade administrators and civil servants into whose hands was put the running of industry, commerce, and agriculture. Their business, as they interpreted it, was to see that the output of workshops and farms was adequate, and that any commercial enterprise such as shipping ran at an acceptable level of profit. These men were responsible for all the technological processes involved in the enterprises they were controlling, yet their recipes for increasing output were seldom other than to employ more labor. Thus, in a large-scale process such as the cupellation of lead ores for the recovery of silver, where, because of lead poisoning, the life of a slave could be reckoned in months rather than years, nothing was done to improve the unhappy lot of the slaves, although the cost of replacing them was enormous and had to be borne against the profits of the process. It was this lower order of administrators whose duty it was to understand the processes of which they were in charge and it was from their ranks that new technological innovations should have come. Instead, they behaved like perfect civil servants: they kept their businesses, their estates and their workshops in good order; asked no intelligent questions; and got, therefore, no intelligent answers.

During the thousand years of which we are speaking, which embraced the height of both Greek and Roman power, there were in fact few respectable occupations. One could be a politician, a philosopher, a jurist, or a general or, preferably, all four. It was also quite acceptable to be an artist, provided one had a sufficient number of slaves to do the dirty work. Under such social conditions it is not difficult to see how the engineer came into his own. He could build a city water supply or drainage system and this would enhance the current politician's reputation. Alternatively, he could devise machinery essential to the better conduct of warfare, which would make him acceptable to the military. To be blunt, because the results of the engineer's handiwork, in itself neither particularly arduous nor dirty, were immediately apparent, the field of

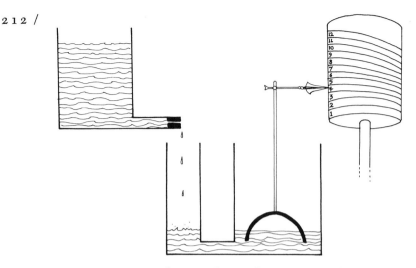

FIG. 205: *Diagram showing the mechanism of Hero's water clock, based on literary sources.*

FIG. 206: *Diagram showing the mechanism of Ctesibius' water organ, based on literary sources.*

FIG. 207: *Diagram showing the mechanism of Ctesibius' fire engine, based on literary sources.*

Most of the machines developed by the scholars of the Museum at Alexandria in the last two centuries before our era are known only from literary sources. The three hydraulic devices shown here are reconstructions. The need to graduate the drum showing the hour in Hero's clock was dictated by the fact that the period from dawn to dusk was subdivided into hours which, therefore, changed in duration with the seasons. Ctesibius' organ worked on the principle of displacing a volume of air by water, while his fire engine—a double-action pump—depended upon the use of clapper valves which were probably already employed in bellows used by metalworkers.

engineering was, as we have said, the only aspect of technology to which an intelligent man could apply himself.

Having been forewarned that this chapter must deal largely with mechanical devices and with building, we now return to the situation in Greece. The Greeks, having been conquered by Philip of Macedon, were in turn carried to conquest in Asia and

FIG. 208: *Diagram of Hero's device for opening temple doors.*

FIG. 209: *Hero's steam turbine, a diagram based on literary sources.*

> Two devices which depended upon the expansion of air or the vaporization of water when heated. The mechanism for opening temple doors when a fire was lit on the altar was typical of the "magical" purposes to which the inventions of the Museum were too often put. Hero's steam turbine could have been further developed to provide a useful source of power. As it was, the machine appears to have been regarded simply as an amusing toy.

Egypt by his son Alexander. At his death in 323 B.C. Alexander had engulfed the whole of the civilized world with the exception of the small colonies at the western end of the Mediterranean. This vast empire was divided among his generals, one of whom, Ptolemy, took as his part Egypt, proclaiming himself king and giving himself the name Soter, or Savior. Despite this exaggerated view of his own importance, Ptolemy established what was essentially a research institute by founding the Museum at Alexandria, of which in time the library was to become the most famous in the world. To the Museum came a great number of scholars both to learn and to teach. Among them, one of the most important was Hero, who had been a pupil of Strato, a contemporary of Aristotle at the Lyceum in Athens. Indeed, it is difficult to say what proportion of Hero's writings depended upon the work of Strato and how much was original. He compiled, nevertheless, what could correctly be called a textbook of engineering. But not all the scholars at Alexandria were imports, for there were local products also, as for example Ctesibius, the son of an Alexandrian barber. Hero, Ctesibius and another Greek contemporary, Philo of Byzantium, were responsible for the invention not only of a considerable number of useful pieces of equipment, but also of several interesting gadgets that could have been put to useful purposes, given the right social atmosphere.

Most of these inventions depended upon a few simple principles including the siphon, the spring, the screw, the pulley, the lever, cogs, cams, valves, and the discovery that air expands when heated. Apart from machinery incorporating these elements, great improvements were made in clocks, in using water as the means of movement, and in astronomical instruments. Hero, for example, invented a water clock and from his written description it is evident that he gave great thought to maintaining an even flow of water into the mechanism so that it would keep accurate time. Ctesibius occupied much of his time devising pieces of artillery and is credited with at least one weapon that worked by compressed air. But alas, the materials of the day were not up to the design and the machine never functioned properly. He had more luck with his fire engine which was, in fact, a double-action pump. Philo also worked extensively on artillery and produced a treatise on ballistics in which he attempted to explain the forces at work when the machines were fired.

Much of the output of these inventors, however, went into creating interesting gimmicky devices often used in temples, such as doors that opened and closed when a fire was lit on the altar. The hot air from the fire caused a volume of air to expand and thus, through a system of gears and pulleys, opened the doors. Ctesibius invented a water organ in which air was drawn through the pipes by a falling column of water. Hero even invented a primitive steam turbine, but it appears to have been put to no mechanical use. Indeed, when one looks at the writings of Philo, which include such topics as the defense and siege of towns, the building of harbors, the principles of the lever and pneumatics, and his work on ballistics, one can appreciate the restricted uses to which the brain power of the Museum was put. Essentially it was for one of three purposes: to beautify the cities, to serve the army or to mystify the worshippers in the temples. At no time does it seem to have crossed the minds of these ingenious men that their inventions could have been used to provide new sources of power or to make industry more efficient, yet they came within a hair's breadth of a real industrial revolution. Many of Hero's inventions, for example, could have been applied to the production of power. He even devised a screw cutter which would have allowed the assembly of really sound machinery. But the failure of the Museum is perhaps summed up best in the piece of artillery worked by

FIG. 210: *Diagram showing the screw press described by Pliny.*

FIG. 211: *Terra-cotta model of a slave working an Archimedes'*
screw pump.

FIG. 212: *A screw pump in use in Egypt today.*

compressed air, invented by Ctesibius; the inventiveness was there, but it was applied to a nonproductive purpose, while the materials available and the means of shaping them were inadequate for the manufacture of so sophisticated a piece of equipment.

The final word, however, must go to Archimedes, undoubtedly the greatest engineer of that age. Born in Syracuse, he is said to have visited the Museum as a young man. His inventions included a screw pump for lifting water from mines, systems of compound pulleys and levers for raising heavy objects, as well as a great number of military engines for the defense of his native city —among which was the proposed use of large, concave mirrors in order to concentrate the sun's rays upon attacking ships to set them afire. As with the others at Alexandria, however, few of his inventions could have been as efficient in practice as they should have been in theory; and yet he gave no thought to improving the materials of which they were made or the manner of their construction. Indeed, when he was asked if he would write a handbook on engineering he refused, giving as his reason that the work

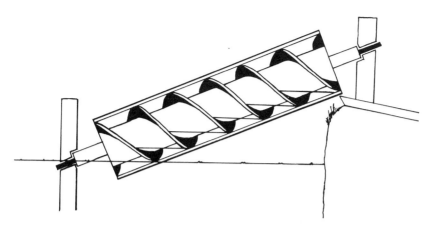

FIG. 213: *Diagram showing the way in which the screw pump functioned.*

A small proportion of the machines developed by the scholars of the Museum and by Archimedes depended upon the use of the screw. It is difficult today to say when and where the screw was first used. It may well be that the screw pump, to which Archimedes' name has been given, was already in use in Egypt before his time and that he merely popularized the machine. In one of his treatises Hero describes a device for cutting screws. The screw press for extracting oil from olives may be seen as typical of the purposes to which the screw was most widely applied.

of an engineer and, indeed, everything that would in any way make life easier, was ignoble and vulgar.

It would be totally wrong, however, to imagine that the inventions of these great men were put to no practical purpose at all. From this time on, cogwheels and pulleys, levers and pumps were widely used throughout the ancient world. Their application certainly simplified a great number of processes, more particularly that of building, where the lifting of heavy stones, for example, became far less of a problem. When we look at the illustrations of machinery of that age, what is so surprising is not so much that it existed at all—for there were plenty of minds to create it—but that it was so clumsy in construction and that the materials used

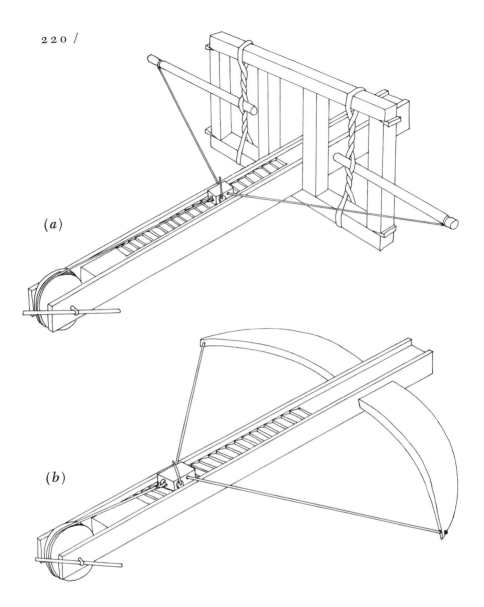

(a)

(b)

were often so inadequate for the function to which they were being put, while to overcome the inadequacies of the materials being used machinery often had an enormous bulk which must frequently have been a trial of strength to move and manipulate.

Unlike Archimedes, the Romans do not seem to have cared unduly whether engineering was vulgar or not, and in their rapid conquest of the civilized world they soon learned to appreciate and to copy other people's mechanical devices. Indeed, the major

FIG. 214: *Reconstructions of the catapult (a) and crossbow (b), based on literary sources, and a photograph of a working-model catapult manufactured by a modern firm of toymakers.*

A considerable part of the work of the Museum scholars was devoted to the design and study of military equipment. The crossbow and the catapult, in which the power was provided by skeins of sinews held under tension, were both later to be improved by the Romans. It is far from clear from written sources how the cocking and trigger mechanisms worked in these weapons. A number of treatises on ballistics show that the students at Alexandria were concerned with increasing both the accuracy and the range of these devices.

contribution of Rome to the advance of technology lay in the ability of its citizens to absorb ideas from elsewhere and to provide an administration that would allow them to be used to their greatest advantage. The layman is probably familiar with many aspects of engineering as practiced by Rome, yet possibly fails to appreciate that virtually every aspect had been invented and practiced elsewhere often long before it was acquired by the Romans. The large public buildings of Rome were inspired by, if not

FIG. 215:
*Roman relief showing a crane in use during
building operations.*

FIG. 216:
*Diagram showing the probable elements of the
Roman crane.* (*opposite*)

The Greek scholars also devoted a great deal of
attention to the study of systems involving pulleys.
Tackles for lifting heavy weights were obviously
much in demand for the construction of large
buildings, as in the case of the Roman crane depicted
here. As a means of providing power the treadwheel
had already been applied to the screw pump, although in a very much smaller form (see Figure 211).

actually designed by, Greeks, and although the Romans appear to
have excelled in the art of building aqueducts (in the first century
A.D. Rome could count nine such structures bringing water to the
city), the Greeks, Assyrians, Babylonians, Persians and Egyptians
had all been building aqueducts for many centuries. The same may
be said of the city's drainage system. Also, the Roman roads, which
crossed the empire from boundary to boundary, were often no
better than the Greek and Persian roads that had preceded them.
Indeed, one author has rather scathingly said that the Roman concept of a road was that of a wall buried in the ground, the top of
which was paved and along which the traffic moved, a view which,
though somewhat exaggerated, embodies a little truth. Roman
roadways were often elaborately heavy when one examines their
intended purpose. Equally, the greater part of Rome's military
equipment, particularly machinery such as the catapult, as we have
just seen, was of Greek invention, whereas her early naval vessels
were timber-by-timber copies of those used by the Greeks and
Carthaginians.

It would, however, be quite unfair to give the impression that the Romans did not appreciate technology even if they were not particularly inventive. Nowhere is the appreciation of what could be achieved by applied technology more clearly seen than in the writings of the Latin encyclopedists. Strabo, for example, who died about A.D. 25, filled his compendious *Geography* with comments on technological processes; although many of his observations were at fault in detail, at least he felt them worth recording. The same can be said of Pliny, born just before the death of Strabo, for a large part of his *Natural History* is occupied by a discussion of raw materials and the techniques used to work them. The Roman attitude to technology was, indeed, subtly different from that of the Greeks. Like the Greeks, the Romans still believed that actually to work with raw materials, to be an artisan, was degrading; nevertheless they did not hold technology in such disdain that they refrained from writing about it. Thus, for example, the Greek philosophers were apt to write about medicine and surgery as abstract studies, while the Romans not only followed them in this but also built hospitals and organized a medical service for their army. To the Greeks mathematics was on the whole an interesting mental exercise, but it was the Romans who used mathematics to build roads and aqueducts and to lay out their somewhat stereotyped city streets. It is to this aspect of Roman genius that we must devote the remainder of this chapter.

One of the most interesting features of the technology of this period was the use made of the mechanical devices that Rome had inherited. These were all based essentially on gears, pulleys, the screw and the lever. Thus we find that the crane used during the erection of large buildings was essentially a sheer-leg tripod, the load being raised by means of compound pulleys, a system that probably differed in no way from that used by the Greeks except that as motive power the Romans attached to it an enormous treadwheel, which itself was possibly developed from the same type of device used when raising water with a screw pump. Elsewhere, instead of the treadwheel, we find the capstan was being used, which in turn was borrowed as a concept from rotary olive and wine presses or from corn mills. By the end of the fourth century A.D. an unknown writer speaking on the subject of warfare goes so far as to describe a paddle-driven ship in which the paddles were manipulated through gears by capstans turned on the deck by yoked oxen. Probably such a ship was never actually built, but it is interesting to see even the theoretical appearance of a paddle wheel at this early period.

Mention of the paddle wheel brings us to one of the more curious aspects of Roman engineering, namely, the spread of the water mill throughout the Roman Empire. One does not know when and where the first water mills were developed, although written evidence shows that they were certainly functioning by the first century B.C. in northern Greece; it was presumably either there or in western Anatolia that they were first put to use. They were, to be precise, water turbines, and they could operate only in an area where fast streams could be channeled to produce a jet of water. The wheel was set horizontally, that is to say, with a vertical axis, and the water was directed onto the blades of the turbine. The axle passed through the lower horizontal millstone to rotate the upper stone—there being no gearing—so that the turbine operated at the same speed as the upper stone. Mills of this type are still to be found in many parts of the world today, but

FIG. 217: *The lower stones of donkey mills, from a building at Ostia.*

FIG. 218: *Diagram showing how the donkey mill was operated.*

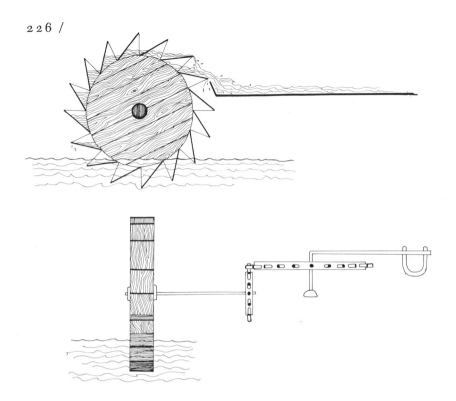

FIG. 219: *Diagram showing how the water-raising wheel operated.*

FIG. 220: *A water-raising wheel in use in Cyprus today.*

FIG. 221: *Diagram of a paddle-driven ship, based on literary sources; probably never built.*

During this period the capstan appeared in many guises. In its simplest form it is to be seen in the rotary mill, in which the upper stone was made to rotate by a horizontal beam to which was yoked a donkey or an ox. With the addition of simple gearing the same principle was applied to the water-raising wheel, first known from Egypt and still to be seen in use there today. The rather complex paddle-driven ship, in which the paddle was worked through gears by ox-drawn capstans on deck, is known only from literary sources, and in all probability was never put to serious use.

only in those areas where the river system will allow the water to be run through narrow channels and ultimately down a steep chute directed onto the blades of the turbine. In lowland areas where streams are sluggish such a mill could not be made to work.

Quite a different type of water mill, first described by Vitruvius in the first century B.C., worked on the principle of the water flowing *beneath* the wheel, which was set vertically, and in so doing striking the blades to make it rotate. The water wheel drove the millstones through a gearing in which the ratio was five turns of the millstone to one turn of the water wheel. Such a water mill obviously had a wider application than the previous type, and furthermore it was capable of doing far more work. Whether or

FIG. 222: *A recent picture of a horizontal mill from Norway.*

FIG. 223: *Diagram showing how the horizontal mill operated.*

> Mechanized milling of corn is known from written
> sources to have been developed by the first century
> B.C. in northern Greece and western Anatolia. The
> type of mill used appears to have been essentially a
> turbine in which water from fast mountain streams
> was directed through a chute on to the blades of the
> water wheel, so rotating the upper millstone. This
> device seems to have spread rapidly throughout the
> Roman Empire, and in basically the same form is to
> be seen in many mountainous areas today from the
> Middle East to northern Europe.

not it was developed from the horizontal water wheel is a matter
of debate, but it should be noticed that it bears a strong resemblance in design to the water-raising wheel in which a draught
animal working through a capstan rotated a wheel bearing cups at
its rim, each cup dipping into the river and being discharged into
a chute after little more than a half-turn of the wheel. There is
reason to believe that water-raising wheels of this sort had been
in use in Egypt for some centuries before our era. It may well be
that the so-called Vitruvian water wheel was developed from
these. Be this as it may, the Vitruvian water wheel had its limitations, for it was dependent for its proper functioning on a level
flow of water in the river: were the river in spate or too low, the
wheel would not function.

By the fifth century A.D. yet a third type of wheel appeared.
This was the overshot wheel in which the water, ponded back in a
mill pond, was directed by means of a chute over the top of the
water wheel, and which thus, by controlling the level of the water
in the mill pond by means of sluices, could be made to function at
most seasons of the year throughout the larger part of the Roman
Empire.

Curiously enough, the Romans seem to have used the water
wheel largely for grinding corn or for similar purposes, such as
extracting oil from olives, although at times great ingenuity was
shown in devising mills. Thus, there is a recorded instance where
a row of wheels was set one above another up an incline, allowing

Drawing of a mosaic from Byzantium, partly reconstructed showing a watermill of the Vitruvian type.

FIG. 225: *Diagram showing how the Vitruvian mill operated.*

Diagram showing the functioning of an overshot mill.

A water mill that could be made to work in regions where rivers were sluggish was first described by Vitruvius in the first century B.C. This was the "undershot" mill, in which the water passed below the wheel, striking the blades to make it rotate. The millstone was made to work through a gear system very similar to that used in the water-raising wheel previously described. It is apparently this type of mill that is illustrated in a mosaic from Byzantium, modern Istanbul.

The Vitruvian mill could operate only when the water in the river was neither too high nor too low. With the later introduction of the "overshot" wheel, in which the water was ponded back in a pool kept at constant level, water mills could be made to operate in nearly all regions of the Roman Empire.

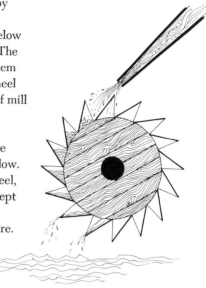

the same volume of water to turn a whole battery of mills. Equally, when Rome was besieged by the Goths in the early part of the sixth century, a floating mill was moored across the Tiber, the flow of the river turning the undershot wheel. But at no time do the Romans appear to have adapted the wheel to any other type of process. There was, for example, no mechanical reason why the water wheel should not have been used to drive bellows or heavy hammers as was done in contemporary China or during the medieval period in Europe.

The Roman city is interesting more for its scale than for the novelty of its design. The aqueducts, the lead delivery pipes and the earthenware piping used for drainage, had all been anticipated elsewhere, although the sheer number and length of Roman aqueducts is in itself impressive. Probably the Romans' greatest contribution to building was the development of the brick-and-concrete arch. This was essentially a brick arch reinforced with a heavy infill of concrete. The arch, once the concrete was set, was essentially a vast lintel and thus exerted very little lateral pres-

FIG. 227: *Diagram showing the stresses involved in a true arch, the Greek pillar and lintel, and the Roman arch.*

FIG. 228: *A view of the interior of a public lavatory, Ostia.*

FIG. 229: *A reconstructed view of aqueducts outside Rome carried on long series of arches.*

One of the few original contributions of Rome to building technology was the introduction of cement. Apart from its use as a bonding material, it was also employed in the making of concrete, which, combined with a brick facing, allowed the construction of solid arches, so eliminating the need for buttressing the supports. Arches of this type were to be seen in their most dramatic form in the building of aqueducts. Indeed, the Romans devoted a great deal of time and money to public sanitation, creating a water supply, drainage and other forms of sanitation often as good as, if not better than, those to be found in many parts of Europe today.

sure, so that the pillars on which it was set required no buttresses to support them. It was a series of such archways that carried the larger part of the raised portions of the aqueducts, many of which still stand to this day. Curiously, for many of their early public buildings the Romans depended upon Greek architects, who introduced a style based almost exclusively on upright pillars capped by lintels, a somewhat cumbersome type of building which, despite its great aesthetic merits, was nevertheless designed to withstand earthquake tremors. However, apart from Rome and the major cities of the Empire, more typically Roman building materials were used and very high technical standards were achieved not only in the minor provincial cities but in the homes and estates of Roman administrators and businessmen. Thus, in Western Europe one need not be surprised to find well-constructed fortifications, city walls and public buildings within the cities. But the high quality of building in the Roman villas, for example, is quite extraordinary and shows that a great deal of money must have gone into their creation.

A number of Roman writers, in fact, deplored the building of these expensive estates, not only in Italy but throughout the whole Empire, arguing, probably quite rightly, that their construction was drawing a lot of wealth away from Rome itself; and we can see in this tendency yet one further cause for the lack of technological development during this period. Large country estates, no matter how well run or how well built, tended to be isolated and their occupants to become self-supporting. Thus, a well-set-up Roman estate would have its own farms and mills, its own ironworks and possibly its own pottery. Indeed, the larger the estate became, the more likely it was to become self-supporting. The technological processes carried out on these estates would nevertheless have been on a comparatively small scale, and provided they supplied the immediate needs of the estate there would have been no impetus to increase their output. Just occasionally, however, we get a glimpse of some unusual development where, perhaps, there was a shortage of manpower. Such an instance may be seen in the development of a reaping machine in Gaul which was described by Pliny. This was essentially a two-wheeled cart pushed by a pair of oxen yoked behind the vehicle. Along the leading edge of the cart was set a sharp steel comb at a height a little below that of the heads of the grain. The stalks of grain were cut by the comb and the heads fell behind into the cart. The machine was probably never very efficient, but when the period during which reaping could take place was short and the work could not possibly be undertaken by the available number of men equipped with sickles or scythes, the device was obviously of enormous advantage. The machine could probably only be used in fields that were sufficiently

FIG. 230: *Low relief from Belgium showing a reaping machine (vallus) at work.*

FIG. 231: *Reconstruction of this type of reaping machine, based on this and similar reliefs and on literary sources.*

In many respects the Romans were slow to appreciate the possibilities of mechanical devices. A simple reaping machine was developed in Gaul, for example, but it was neither developed further nor did it spread to other parts of the Empire, despite the fact that it could have greatly eased the shortage of manpower that crippled the later centuries of Roman rule.

large and level, but it is again symptomatic of the whole of the Roman period that it was not developed to become more efficient and did not achieve more widespread use.

Apart from the development of her luxurious roads, little progress was made by Rome in the field of transport, and although we know of a number of light vehicles in which the draught animals were harnessed between a pair of shafts and in which the normal yoke was replaced by a padded collar, for moving heavy burdens the older, paired draught and yoke were still used. Although the Romans were well aware of the potentialities of proper stock breeding, they appear to have ignored the possibility of rearing a line of heavy draught horses, just as they failed to develop shafts and collars for the animals pulling their heavier wagons.

The same static position is to be seen in Roman sea and river transport. So far as their war galleys were concerned, they borrowed from the Greeks and Carthaginians the low vessel with a ram, the oars now being rowed through decked outriggers which had the advantage of giving the crew a greater leverage while bringing the oars into a straight line, while at the same time the decking gave the oarsmen some protection from the enemies' missiles. Even by the fifth century A.D. Roman merchant shipping was much as it had been inherited from the Greeks: fairly short vessels high at both bow and stern. At some point, probably in the first century B.C., a small, square jibsail was adopted in addition to the square mainsail, but the vessels could still only be worked in a following wind. This severely limited the possibilities of sea commerce, since the cost of rowing merchant vessels for anything other than short distances was financially prohibitive. In the field of shipping, however, the lack of progress during the period of Roman domination is perfectly intelligible, for Rome had no competitors on the seas in which she did business other than pirates, whom her men-of-war were always capable of destroying.

Perhaps the most remarkable feature of the Roman period is the degree of sheer sophistication accompanied by utter crudity in much of the technology, even in essentially the same fields. Thus, for example, the Romans invented the most elaborate heating systems, not only the type of under-floor heating, the hypocaust, well known to most people, but also boilers with plumbed

FIG. 232: *A Roman galley as depicted on a mosaic from Palestrina*
(*Praeneste*).

FIG. 233: *Reconstruction of the bow and profile of a Greek ship
of about 200 B.C., showing the decked outriggers
projecting beyond the ship's sides; based upon a badly
damaged marble statue, vases and low reliefs.*

FIG. 234: *Roman merchant ships, as shown on a mosaic from Ostia.*

Few changes were made in the design or construction of shipping under Rome. The war galley remained much as the Romans found it in Greece and Carthage. The three-bank ship seems to have been abandoned in favor of a two-tier system in which more than one man was set to each oar. Rowing was invariably carried out through decked outriggers, so protecting the crews and bringing the blades of all the oars into a straight line. Outriggers were certainly in use in Greece by the beginning of the third century B.C.

Merchant ships were often fitted with small square jibsails, but otherwise the hull and general upperworks were very similar to those in Greece in the fourth century B.C.

FIG. 235: *Copies of Roman and Old Stone Age lamps.*

FIG. 236: *A Roman "vacuum" flask from Pompeii, made of bronze and iron.*

Roman technology was riddled with strange contrasts. Their engineers were quite capable of installing plumbed hot water or of devising a "vacuum" flask to keep food warm. On the other hand the normal form of lighting was a small oil lamp, a refined version, but functionally no different from lamps used in the caves of southern France during the last Ice Age.

FIG. 237: *A Roman "air bed": a woodcut of the sixteenth century based on Roman literary sources.*

The inflated mattress, made of leather, was never widely adopted, if indeed ever used at all. This was a brilliant idea, but as with so many inventions of the period the materials of which it was made were hardly suitable.

hot water available on tap; while their fundamental grasp of insulation allowed them to construct a primitive form of thermos flask. Yet when it came to lighting, the lamp, which was no more than a small covered bowl with a circular hole through which the wick protruded, was barely any advance on that used by the artists who painted animals during the Ice Age deep in the caves of southern France. Indeed, if one wished to be altogether cynical about the matter one could say that the only advance made was in the cover over the lamp, which was not infrequently decorated with obscene reliefs. Again, we know from written sources that the Roman ideas of anatomy, physiology and medicine were extremely unscientific and that the proper functions of many parts of the body were unknown; yet surviving surgical instruments show us not only that the Romans were able to carry out a number of reasonably complex operations, but also that their instruments were extremely well designed for the work. Many of their obstetric instruments, for example, can be paralleled almost in detail with those in use at the beginning of the present century.

When the end came, the Roman Empire was overrun by barbarian tribesmen from northern Europe and South Asia. In this final defeat some writers have seen a technological failure on the part of Rome: a failure in that she was unable to meet, despite her higher technology, the threat of these barbaric people. Some historians have gone as far as to say that these tribal horsemen, now equipped with the stirrup, were such an effective cavalry that the Roman legions were quite unable to withstand them. This view fails to take into consideration the fact that for nearly three centuries the Romans had recruited many of their troops from among the barbarians and that both barbarians and Romans were well aware of the equipment and methods of fighting adopted by each other. It fails equally to take into account the fact that well-mounted and well-equipped cavalry in the middle of the last century in the United States was not infrequently outmatched and utterly routed by poorly equipped Indian tribesmen riding with neither saddle nor stirrup.

Before we pass final judgment on this question we should examine in rather more detail the state of technological advance of the barbarian world.

8/The Barbarians of Northern Europe, India, China and the New World

THE BARBARIANS IN THE WEST

To the writers of classical antiquity, anyone who did not share in one of the great civilizations of the Mediterranean world was a barbarian. The impression that they would have us gain is that such people were uncouth, untutored and thoroughly backward in every respect. And yet an author such as Caesar could barely conceal his surprise and admiration for the Gauls in the seaworthiness and sophistication of the ships they sailed in the Western Sea. He was forced to admit, for example, that their vessels were more sturdily constructed and better designed to withstand the weather of the Atlantic than those of the Romans. He even noted that their anchors were secured by chains and not by cordage. In effect, what Caesar said was that though these people might have been barbarians their shipping was excellent. However, when he spoke of Britain, he claimed that the inhabitants of the interior did not grow grains and dressed themselves in skins. But then, Caesar had never set foot in the center of Britain to see for himself what the situation was, and like so many authors of his day he was as much a dupe to propaganda as he was himself a propagandist. Sadly, it must be admitted that the writers of classical antiquity are not the most reliable sources if we wish to know the state of technological advance of the barbarian world.

This attitude of a higher civilization to a lower is, of course, still with us today. "No peasant ever invented anything," is the rash statement of one eminent anthropologist. Even on the face of it this is an absurd statement, although it seems to embody an element of truth. One can ignore the statement at face value and assume that what is really meant is that peasants are less inventive than city dwellers. We might do well, therefore, to begin by deciding whether the barbarians were really less inventive than their more civilized neighbors to the south and east.

We have already spoken of the shipping of the Gauls, which was so sturdily built that the Roman warships found it almost impossible to sink by ramming. These strong vessels, which Caesar used extensively later on as transports, were hardly the product of an unimaginative people. Equally, while Roman oxcarts creaked their way along their sophisticated roads, a simple carpenter in northern Europe who had probably never seen a Roman, and who was certainly illiterate, designed and built a wheel that moved on roller bearings. In this context it matters little that his invention probably died with him and that he chose the wrong material of which to make this clever device. Yet another peasant, this time

FIG. 238: *Diagram showing the structure of a roller bearing from the wheels of a cart from Denmark; about 100 B.C. The device was made of wood and bronze.*

The craftsmen of barbarian Europe were perfectly capable of designing and constructing quite sophisticated pieces of equipment. This roller bearing is a good example of the inventiveness of the people whom their more civilized neighbors looked upon as savages.

FIG. 239: *Drawing of a bronze-bladed sickle with a wooden handle, from Switzerland; about 1000 B.C.*

The design of tools and weapons was often as sophisticated amongst the barbarians as in the centers of civilization themselves, as may be seen by the form of this sickle handle, carefully shaped to fit and to protect the reaper's hand.

FIG. 240: *Diagram showing the mechanism of a prehistoric wooden trap for catching deer.*

FIG. 241: *Drawing of a low relief from an early Christian cross,*
Clonmacnoise, Ireland.

> Many of the devices in use in barbarian Europe
> incorporated features that could, at least in theory,
> have been put to other uses. Although traps of this
> kind were in common use throughout Europe, the
> valve, kept in the closed position by a springy stick,
> seems to have been used in no other context.

living in the Alps, designed a sickle handle to fit precisely the
contours of the hand. Again, his invention appears to have died
with him, although today it would undoubtedly win an award at
any design center. Once more, those same peasants whom Caesar
believed wore skins invented a very simple beater-in: a comb with
short teeth with which to compact the weft in the textiles they
were weaving, the same small tool that occurred universally
throughout Britain but nowhere else in Western Europe. These
few examples, which could be multiplied many times, should be
sufficient to dispel any preconceived ideas about the lack of inventiveness of the prehistoric peasant.

There existed, furthermore, among the barbarians a number
of fairly sophisticated pieces of equipment, facets of which we
have already noted in our review of the main stream of technological development in the Near East. Thus throughout a large

FIG. 242: *Bronze shield from the Thames; late first century* B.C.

Metal objects, although never attaining quite the
dimensions of those of the civilized world, were quite
as sophisticated and as intricate among the barbarians.

part of Western Europe there was in widespread use a wooden trap with a spring-loaded valve designed to catch deer. The valve was, in fact, a small hinged door let into a slot cut in a log, and kept in the closed position by a length of springy wood. The whole device was buried in the pathway used by wild deer, and when the animal put its foot on the valve the spring released it sufficiently to allow the foot to sink through the hole in the log where it was held firm by the valve under the tension of the spring. With very little adaptation the makers of these traps could have put this valve to many other purposes had the need arisen, and we cannot even say that the barbarians lacked the necessary mechanical skill with which to make such technological advance. Indeed, one of the most interesting aspects of the study of prehistoric Europe is the way in which various technologies were acquired from the higher civilizations of the Mediterranean and applied by these people to their way of life. What is perhaps yet more interesting is not that many technologies were acquired at all, but the selection of those that were and those that were not adopted. At first sight there seems to be little rhyme or reason in the choice of materials and techniques taken over by the barbarians from the more civilized world.

In the long period prior to the use of metals the peasants of prehistoric Europe gradually acquired cultivated crops and domesticated animals, as well as many crafts that one would expect to find associated with these, such as the use of millstones, the making of pottery and fairly complex woodworking based upon the use of stone tools, one aspect of which was the building of elaborate barns and houses. But here the analogy ends. As far as we can judge, many of their villages were of a completely temporary nature and many more were occupied for only a short period of time. The villages, too, were minute by Near Eastern standards, and nothing even approaching a small town seems to have come into being.

Later, these same people were to acquire the use of metals, gold and copper in the first instance, and then bronze, with extraction and casting techniques every bit as sophisticated as those to be found in the Near East. But glassworking was quite unknown to them, and their pottery by Near Eastern standards remained for a long period extremely crude. Indeed, during the Bronze Age a most extraordinary situation was to be found, in which clay molds used for the manufacture of bronze implements were made with the greatest of care, considerable attention being given to the nature of the raw materials used in their manufacture. At the same time the pottery used by the same people was unspeakable rubbish, thus showing that while they were able to employ ceramic materials intelligently they did not always choose to do so. Accompanying the spread of the use of metals we also find the building of carts and plows, but the chariot by contrast was slow in its diffusion throughout the prehistoric period in Europe, and for the larger part its appearance can be equated with that of ironworking. Villages remained small and still of a rather impermanent nature, while the prehistoric barbarians knew nothing of writing and recording either on clay or on any material at hand.

The prehistoric Europe into which Rome eventually moved, therefore, was one in which iron technology had spread, in which chariots were still in use as a weapon, at least in Britain, and in which the people used gold coins. The barbarians had learned to make pottery on the wheel, and to turn wood on a lathe. In one place they had even made an experiment in building in a totally inappropriate material, mud brick, a method of construction that they seem to have abandoned equally rapidly. The Gauls and their neighbors were, nevertheless, an illiterate people living in villages. We are probably not far from the truth if we assume that the ability to read and write went hand in glove with city life and that all that was really missing in prehistoric Europe was the appearance of an urban civilization. We can even narrow the argument further to a discussion of why the city never took root in barbarian Europe.

For a city to come into being demanded a certain degree of social organization, and it might be thought possible that in prehistoric Europe such an essential organization did not exist among the rather scattered tribes that occupied the territory. But this cannot be totally true. There have survived even to this day a sufficiently large number of elaborately constructed tombs, meeting places and fortresses, as well as the occasional temple, such as Stonehenge, to show that where there was a will the population

FIG. 243: *A collection of bronze scrap metal from Saxony; about 700 B.C.*

The barbarian bronzefounder was essentially an itinerant, setting up a temporary workshop as and where required. Dumps of scrap metal, previously collected from customers and hidden for future use, have often been discovered, normally unrelated to buildings or settlements of this period.

could be organized in sufficiently large numbers to carry out this type of work. The elaborate banks and ditches around the central enclosure of many fortresses, for example, could not have been carried out without a strong central direction and a very considerable labor force.

Perhaps we will understand the nondevelopment of the city in prehistoric Europe better if we look briefly once again at the development of the city in the Near East. By many writers this development has been described as essentially the growth of a complex of dwellings around a central temple, shrine or place of worship; and this indeed is often the superficial appearance given by such cities upon excavation. But in fact most cities were primarily the centers of particular technologies. Throughout the whole period covered by this book most cities were renowned simply for the presence within their walls of one or two main technologies such as the making of pottery or glassware, or for the manufacture of jewelry, and there is very good reason to believe that more often than not it was these industries, and not the presence of the temple, that brought the city into being. Of course, where a technology developed there had to be a means for the import of its raw materials and the export of its finished products so that the city could become equally a center of trade and a market place for the buying and selling of agricultural products and livestock to provide food for its citizens. This argument must, therefore, lead us to the point of view that in some degree the main distinction between technological development in the Near East and in prehistoric Europe was that the barbarians failed to centralize their industries.

To a limited extent this point of view is borne out by the nature of those technologies absorbed by the barbarians and of those which they apparently rejected. For example, it was not necessary to construct an elaborate foundry for the production of bronze weapons and tools. The bronzesmith himself might have been completely mobile, shifting from one community to another, on each occasion setting up a temporary furnace to produce the objects required. The fact that dumps of broken bronze objects, clearly intended to be resmelted, have frequently been found unrelated to any building shows that in prehistoric Europe the bronze industry was operated largely in this way. On the other hand, glassworking did require a more permanent setup; although just possible, it would have been extremely difficult for a glassworker to move all his equipment from one community to another regularly throughout the seasons of the year. Also, as we have seen, the manufacture of crude pottery could have been carried out with few or no large tools at all, whereas the production of sophisticated pottery demanded a lot of permanent equipment, such as wheels on which to throw the pottery and kilns in which to fire it. In the absence of urban centers from which its products could be disseminated, we would, therefore, not expect to find wheel-thrown pottery.

Even in those areas from which raw materials were acquired in bulk, as for example around copper- and ironworkings, salt mines or tin mines, no cities seem to have developed; it would therefore appear that in barbarian Europe raw materials were acquired and distributed in the same *ad hoc* way as that in which the craftsmen moved from one community to another. Yet the volume of trade was often sufficiently large to warrant urban development, and at times it would seem almost as though there were among the prehistoric peoples of Europe a deliberate attitude of thought that ran counter to building cities. In Britain, for example, the Roman policy of creating towns was a signal failure: the local population flatly refused to live in them, and often cities that were planned were never completed for lack of occupants.

Whatever the cause, and it may in the end have been merely that the population in barbarian Europe was insufficient to support cities, their failure to produce urban communities can be seen as a primary cause for their failure to produce or to adopt new techniques. Our past review of the history of technology in the ancient world has shown the enormous reliance of one technology upon another; in scattered communities, where there could have been little or no exchange of ideas, new technologies are likely to have fared poorly and inventions to have died stillborn.

THE BARBARIANS IN THE EAST: THE INDUS VALLEY

So far we have failed to record adequately the fact that the city settlements in the valleys of Mesopotamia and of the Nile had their parallels in the Indus valley. Indeed the archaeological record shows that farming communities, similar in general pattern to those that we have seen in the Near East, had taken root in this part of northwest India well before the year 3000 B.C. Later a number of cities were to develop of which Harappa and Mohenjo-Daro were the largest and are probably the best known. In many respects these cities and the technologies practiced within them were a reflection of what we have already seen in the Near East. Building in mud and later in fired brick; the development of a system of irrigation; the smelting of copper and then bronze; the use of Egyptian faïence and of stamped seals all appeared in much the same order as in Egypt and in Mesopotamia, while in many other respects the civilization of the Indus valley followed very closely that of Mesopotamia. The cubit and the foot, for example, of identical magnitude to those used in Mesopotamia, were the current measurements of length. One might even be excused for supposing that the people of the Indus valley were not particularly inventive and borrowed heavily from their distant neighbors in the West, and that the Indus valley was a cul-de-sac which accepted ideas but produced no new technological innovations itself. This point of view, however, needs very considerable modification for they were not totally bereft of their own technologies.

The people of the Indus valley certainly cultivated new crops and domesticated other animals than those found in the Near East. Shortly after 3000 B.C. cotton was being cultivated as a fiber plant for making textiles, while the humped cattle of India, for which it must be said that no wild prototype is immediately apparent, were in domestic use at an equally early date. The fact that their origin is obscure suggests that stock breeding had been practiced in the Indus valley for a long time before this. While the people of the Indus valley bred all the farm animals common at this time in Mesopotamia, they had also domesticated the water buffalo and tamed the elephant.

If the use of stamped seals and the basis of linear measurement were borrowings from Mesopotamia, the same cannot be said of their script, although of course it is unlikely that the concept of writing was an independent invention of the people of the Indus valley. The script itself is totally different from that of Mesopotamia or Egypt and it remains largely undeciphered. Although clearly based on some form of pictogram, the symbols as we know them are a long way removed from the type of pictogram that illustrates some common object. It is just possible, of course, that earlier forms of this script are yet to be discovered, but at the moment it looks very much as if this form of writing was evolved fairly rapidly to meet an immediate demand. In these circum-

FIG. 244: *Two seals from Mohenjo-Daro; about 2000 B.C.*

In the third millennium B.C. cities developed in the Indus valley with technologies very similar to those of contemporary Mesopotamia. In India, however, the local fauna was exploited, as for example the elephant and humped cattle.

The form of writing seen on these seals (still largely undeciphered) was clearly based on earlier pictograms. Its relationship with that of Mesopotamia, if any, remains obscure.

stances we must suspect that the concept of writing was lifted wholesale from Mesopotamia, and that in the Indus there were no early evolutionary steps.

Whatever the connections, however, between Mesopotamia and the Indus valley they can never have been very strong. Admittedly, seals and other small objects of Indian origin have turned up in Mesopotamian cities, and vice versa, but there are nevertheless extraordinary lacunae in the development of many technologies in the Indus valley that could never have existed had there been strong connections between the two areas. Thus, for example, we find in the field of metal technology first the use of copper and then at a later date the manufacture of alloys, a development strictly comparable with that in Mesopotamia. This being the case, one might expect to see the same sequence of events in the actual design of molds into which to cast the metal, but curiously enough the open mold continued in use over a very long period. It was not until almost the dramatic end of this civilization that piece molds were introduced, and even then they were undoubtedly brought in by intruders from the north.

More than one archaeologist writing about these early cities in the Indus valley has commented upon the dreariness, the repetitiveness and the dead-level nature of the various objects produced by the Indus craftsmen. Looking at the way in which the cities were planned, with their rows of mean huts for the workers, grouped close to furnaces for metal production or to pottery kilns, one has the uncomfortable feeling that here were city-states in which production was ruthlessly organized but in which the techniques employed were probably not very efficient. In fact, one gets the feeling that here again the dead hand of the civil servant was in operation, such as was surmised during the declining years of Rome.

In discussing barbarian Europe, it was suggested that the failure of the primitive peoples of prehistoric Europe to produce new technologies was due to a lack of cities. If the Indus valley civilization is to tell us anything, it perhaps illustrates the fact that cities alone are insufficient for technologies to be thriving and progressive. For the great cities of Mohenjo-Daro and Harappa were as large and as industrious as any known from the ancient world at this date; and while we may blame the way in which their industries were organized, in all probability the root of the trouble lay in the difficult communications between the Indus valley and the rest of the civilized world. As it is, it would seem that these two great cities and their dependent towns and villages became technologically introspective. They had achieved a level of technological development that was satisfactory from their point of view, and there appears to have been inadequate contact with the rest of the civilized world to act as a stimulus to further development.

Many of the achievements of the Indus civilization were abruptly destroyed sometime shortly after 2000 B.C., when northern India was overrun by groups of people from the north akin to those who, as we saw in an earlier chapter, overran Greece and Anatolia. These tribesmen were equipped with superior weapons, including the chariot: the cities of the Indus were destroyed as, too, was their administration. It was left to the invaders to choose which of the existing technologies that they found in the Indus valley they would use.

So far as the history of technology in the ancient world is concerned, the contribution from northern India was to be negligible. It was not until after more than a thousand years that one of the Indus valley's independent developments, cultivated cotton, was imported first into Mesopotamia and then Egypt.

THE BARBARIANS IN THE EAST: CHINA

OUR UNDERSTANDING of the development of civilization in early China is today extremely fragmentary, due almost entirely to a lack of archaeological excavation designed to throw light upon the period of its origins. Although much has been done in the last few decades to remedy this situation, it is still almost impossible to write with conviction about the earlier periods covered by this history. Thus, for example, almost nothing is known of how agriculture and stock rearing came to be practiced in early China. But within the few centuries immediately before 2000 B.C. there were clearly a number of agricultural settlements in the main river valleys, and while we know very little indeed about the economy of these societies, two aspects of their technology—or what remains to us of their technology—are of considerable interest. On the one hand we find that these people were making pottery by the same techniques that had been used in western Asia, but the style of the wares they were producing was particularly their own as, too, were the forms of decoration. The pottery was made on a turntable as it had been in western Asia; it was decorated with pigmented clays painted onto the surface; and it appears to have been fired in some sort of primitive kiln. There is, however, nothing in the Chinese archaeological record to suggest that the art of making pottery was a local development, and we may suppose, for want of evidence, that the technique was initially acquired from the western Asiatic societies that we have already discussed.

A second aspect of Chinese technology at this early date is of particular interest not so much because of the techniques being employed as because of the raw material used, for among the objects from these early agricultural settlements are a number of plaques made of jade. This is a hard rock and difficult to work without metal tools: it is also a mineral of limited origin. It does not occur as a deposit within China proper, and although it has been argued that the stone used was in the form of geological erratics, the argument carries little conviction. The most likely source, in fact, would seem to have been many hundreds of miles from China proper, in the mountains of Sinkiang. The exploitation of a rock so far from its source by a simple agricultural community is, of course, remarkable; but the fact that this was so serves as a clue to the entire problem of the early technology of China. The trade route, stretching across Asia from China to the cities of the Near East, three thousand years later to be known as the "Silk Road," must to some limited degree already have been in use. We are left speculating by whom and for what reward jade was carried this vast distance into China at so early a period.

What has been said about the appearance of the technique of potting in China is equally true of the practice of bronze founding, for about the year 1500 B.C. bronze appeared, so as to speak, out of the blue. There have been found no early attempts at metallurgy, no unalloyed copper tools, none of the simple forms of stone mold that we have learned to expect in the West. Instead one finds that the Chinese were suddenly producing bronze vessels of very considerable complexity, using elaborate methods of casting. We must suppose, therefore, that the technique of casting, the knowledge of how to win the metals from their ores, and to form alloys, were introduced from without, presumably by the same route as that which brought jade. Even so, from the very beginning the Chinese approach to bronzeworking was totally different from that of western Asia.

In western Asia early metallurgists treated their metals as though they were some form of superior stone, and although they were cast into molds to produce the approximate shape of the tool

FIG. 245: *Reconstruction of the kind of Chinese mold used for casting a bronze vessel of about 1500 B.C., based on existing fragments of molds.*

FIG. 246: *Chinese bronze and pottery vessels; about 1500 B.C.*

or ornament required, most of the finishing was done later by hammering and polishing; in very few cases until the later periods does one find molds designed so that little or nothing needed to be done to the surface of the bronze after it had been cast. By contrast the Chinese attitude toward the metal was that of the potter: it was treated like some superior ceramic. The Chinese spent most of their effort in creating molds that were so precise in every detail that, once cast, the bronze object needed little or no fettling. In order to achieve these results, the Chinese metallurgists found it necessary to raise the lead content of their bronzes, and it is not at all uncommon to find the lead content of the alloy as high as fifteen per cent, while on occasion this proportion rose to thirty per cent.

The precise method by which early Chinese molds were made has been the subject of heated debates, some scholars claiming that the whole production was achieved by the use of a great number of extremely carefully fitted pieces into which had been impressed all the details of the objects to be made. Such piece molds might contain thirty or more sections, all of which would have had to be keyed together before the casting was made. Indeed, a number of fragments of such molds have been found, which alone lends support to the view that this was the technique employed. On the other hand, some scholars, after examining surviving bronzes, have come to the conclusion that they could have been made only by the

FIG. 247: *Crucible used for casting bronze, from China; about 1500 B.C.*

In China no early stages of the development of metallurgy are known: the appearance of bronze seems to have been quite abrupt. The techniques of working the metal, however, differed from those in contemporary western Asia, as, too, did the purpose to which it was put. In China the metal was cast into very complex and accurately made molds so that a minimal amount of hammerwork was required in its finishing. A further contrast may be seen in the type of crucible used, which was made with a double wall to insulate the wooden tongs used to grasp it. Curiously, the Chinese of this period do not appear to have made pottery in molds, despite the fact that bronze and clay vessels were often remarkably similar in form.

lost-wax method. On the face of things it would seem probable that both methods of production were being used concurrently, although the piece mold seems to have been the more common of the two; it may well be that those who adhere to the view that many of the bronzes were cast by the lost-wax method have failed to take into account the enormous strides made by the Chinese in the field of ceramics, and have underestimated the ability of the Chinese to make such intricate molds at this early period.

At the same time that bronze was first used in China, very

considerable advances had been made in the field of pottery production, and the local craftsmen were employing one of those rare mineral deposits, china clay or kaolin, in the manufacture of pure white wares. The method of decoration applied to this pottery is curious, for far from being cast, as was the decoration of the bronze vessels, it was carved into the surface after the wares had been allowed to dry, but before firing. This is the more curious since the motifs used on both the bronze and pottery vessels, although seldom identical, are often similar. Indeed, the whole style of decoration is reminiscent of flat carving such as might have been carried out on the surface of a vessel made of bamboo, wood, or even bone. Apart from the early exploitation of china clay, feldspar, previously ground to a fine powder, was being used to form a blotchy and un-

FIG. 248: *Reconstruction of a Chinese chariot of about 1000 B.C., based upon excavated examples.*

The Chinese appear to have acquired the chariot from their nomadic neighbors at much the same time that they borrowed the composite bow, although both seem to have been adopted later in China than in western Asia. The earliest Chinese chariots were fitted with multispoked wheels, but the yoke and other fittings differed little from those in the West.

even glaze. Later it was to become the basis of glazes applied to Chinese stoneware. Although we know nearly nothing of the kilns of these early potters, it is quite evident that they must have been superior in many ways to those being used in western Asia, capable of achieving rather higher temperatures and of greater control of the atmosphere within them. For some unexplained reason neither the white wares nor the feldspathic glazes remained in vogue for long, and by 1000 B.C. both had become uncommon and were shortly to disappear, apparently not to be made again for nearly another millennium.

Bronze, however, was not the sole borrowing from the West, for during this period the Chinese adopted two important weapons: the chariot and the composite bow. The chariot, unlike those we have seen in western Asia, retained its axle in the middle of the platform but the wheels were often fitted with as many as eighteen spokes. It was, nevertheless, pulled, as in the West, by horses in paired draught using the yoke in the form of an inverted

FIG. 249:
Reconstruction of the Chinese composite bow of about 1000 B.C., based on surviving fragments.

The composite bow, made of lamellae of wood, or of wood and horn, seems to have been developed by the Asiatic nomads. A short weapon, but nevertheless with a considerable range, the composite bow was ideally suited for use from a chariot or from horseback, where a longer bow would have been too cumbersome.

letter Y. The composite bow, made of separate lamellae of bone, wood and horn, although comparatively short and therefore easily handled from a chariot, provided as much power as the longer wooden bow of the typical infantryman. It, too, appeared in western Asia rather earlier than in China.

If bronze, the chariot and the composite bow all appear to have been importations from the West, the same cannot be said of early forms of writing among the Chinese. Certainly the first characters of Chinese writing were pictograms, as they were in the West, but they were not used for the purpose of keeping records, nor were they written on clay tablets or on papyrus. They were, in fact, made on bones with the purpose of divination. It would seem that no one of any social standing would have dreamed of starting

FIG. 250: *An early pictogram of a plow from China, of about 1500 B.C., and a drawing made from a low relief of about 200 B.C.*

Early forms of plow in China differed in design, but not in function, from those in use in western Asia. The earliest forms of writing known in China were, as in the West, pictograms, but the symbols were engraved upon bones which were used to foretell the future. The later Chinese brush-and-stick symbols were derived from the pictograms.

a new enterprise without first consulting a soothsayer. His intention was inscribed upon the bone, which was then prodded with the red-hot point of a bronze tool, and depending on the way in which the bone cracked, it was decided whether the omens were good or bad. Originally there were some five thousand symbols, of which we can now only decipher about a third. This is largely due to a complete reform of the system of writing in China in the second century B.C., but those pictograms that we can understand formed the basis of the later script.

Although there seem to be many points of similarity between the development of technology in China and in western Asia during this period, there was a very marked difference in the organization of workshops and industry generally. Broadly speaking, in China the whole of industrial output was in the hands of the rulers, and while in the West bronze was used for tools as well as for weapons, in China almost the entire output of the early workshops went into the manufacture of ritual bronze vessels and, of course, weapons. The farmer, the woodworker and indeed most other craftsmen still had to rely upon stone tools. It was not until the appearance of iron that the ordinary artisan was able to acquire metal tools.

When eventually the Chinese took to using iron, the entire process of its preparation was totally different from that in western Asia, for the earliest iron objects that we know of, from the fourth century B.C., were made not of wrought metal but of cast iron. It will be remembered that in the West iron ore was first reduced to a bloom, and that this was then hammered into wrought iron on an anvil. In China, by contrast, the ore was reduced directly to the molten metal, which was poured into prepared molds, thus directly continuing the long tradition of bronzecasting. This was possible only because of a number of local circumstances. First of all, the majority of the iron ores available in China had a high phosphorus content which greatly reduced the melting point of the metal; second, there was readily available good refractory clay from which to build furnaces able to withstand the high temperatures involved; and third, the Chinese had developed an ingenious form of piston bellows that allowed them to produce a fairly steady draught through the furnace. The cast iron so produced, however,

FIG. 251: *Chinese tools of cast iron with reconstructed wooden handles and blades; about 400 B.C.*

Before the use of iron, the Chinese craftsmen had to rely entirely on tools of wood, stone and bone, for bronze was reserved exclusively for ritual vessels and for weapons. The earliest iron used was the cast metal, which was too brittle for making weapons but could be used for tools. Normally, cast-iron blades were employed as a cladding to existing forms of wooden implements.

was far from being universally serviceable, for it was far too brittle for the manufacture of arms and weapons, and was therefore used largely for vessels and for agricultural tools. It was two centuries before the Chinese learned how to treat their cast iron to produce a malleable form of metal that could be used for weapons.

FIG. 252: *Reconstruction of an early Chinese crossbow, based on a drawing of the fourth century* A.D.

FIG. 253: *Diagram showing the cocking and release mechanism of a Chinese crossbow, based on a surviving example.*

The crossbow was developed in China in the third century B.C. as a measure of defense against the ever-present hordes of highly mobile nomads. This weapon, with a far greater range than that of the composite bow, originally had a single bow spring, later to be made in the multiple form illustrated here. The efficiency of the crossbow depended entirely upon the cocking and trigger mechanism, without which loading and firing would have been fatally prolonged. The cocking device in only slightly modified form is to be seen in many firearms today—including children's cap pistols.

During this period the Chinese were under constant pressure from the nomads on their frontiers, and the incessant incursions of hordes of mounted men armed with the composite bow caused enormous havoc. In response to this threat the Chinese developed the crossbow which, because it could be held steady and had a greater fire power than the weapons of their adversaries, more than evened up the odds for the Chinese. Undoubtedly the most ingenious part of the crossbow was the cocking and trigger device, without which it could never have been an efficient weapon. It is perhaps of interest to note that this same mechanism is to be seen almost unaltered in hammer-operated guns of very much later periods and, indeed, as a release mechanism in many early clocks.

At much the same time in China appeared true draught harness for horses—shafts, traces and padded collars—all of which allowed a more efficient use of these beasts of burden, eliminated the need for paired draught and, because there was no further risk of half throttling the horse, gave chariots and carts a wider field

FIG. 254: *Diagram showing the probable working of Chinese lantern bellows of the second century* B.C., *based on a low relief.*

of operation. It is, however, far from clear whether this development was truly Chinese; it seems more probable on the face of it that here, again, was a direct borrowing from the nomads.

In the second century B.C. new impetus was given to the iron industry by the introduction of a double-action piston bellows permitting a very regular draught through the furnace. It also allowed "fining" of cast iron, a process in which the molten metal was submitted to a steady flow of air, thus removing much of the carbon which had made it brittle, to provide a form of malleable iron that could be used for the manufacture of weapons. Much of the hard work was taken out of the actual process of forging by the use of the triphammer. This had probably already existed as a form of pounder for crushing grain, being little more than a foot-operated pivoted lever with a heavy hammer mounted at the far end.

The same period saw the introduction of a number of related pieces of equipment that were to prove very important in the future development of Chinese technology. The lap wheel, oper-

FIG. 255: *Diagram showing the working of Chinese double-action piston bellows of the second century B.C., reconstructed from documentary sources.*

Efficient ironworking was only possible with improved bellows, which would maintain a steady draught through the furnace. Early forms of bellows from China seem to have been similar to a modern folding lantern, the draught being created by compacting the bellows. Later, with the introduction of the double-action piston bellows, an even more regular draught could be supplied, so eventually permitting the manufacture of steel from cast iron.

FIG. 256: *Diagram showing two types of mill in use in China before the beginning of our era, both based on documentary sources.*

Early Chinese mills were often similar to those found in western Asia during the same period; the same devices may still be seen in use in many parts of the world today. The foot-operated hammer mill, seen here on the left, was possibly what inspired the iron-forging hammer, ultimately driven by a cam on the shaft of a water wheel. This allowed the making of larger forgings that would have been possible with hand-held hammers.

FIG. 257: *Sectional drawing of a Chinese seismograph of bronze; probably about* A.D. 600.

This device for detecting earth tremors may be seen as typical of Chinese mechanical inventiveness of this period. Seen in the round, the vessel carried on its perimeter a dozen figures of frogs, each with a metal ball held in its hinged jaw. Earth movements would have disturbed the central pendulum sufficiently to release a metal ball, thus showing the direction of the tremor.

FIG. 258: *Making paper in China, a reconstruction based on a drawing of about* A.D. 500.

Paper was first made in China about A.D. 100.
From a prepared mash of plant fibers a small quantity
was shaken into an even layer on the mesh of a sieve.
After draining, the paper was removed and stuck
against a wall to dry. For centuries before this the
Asiatic nomads had been making felt from animal·
fibers. It is debatable to what extent papermaking was
inspired by the process of felt manufacture.

ated in a very similar manner to the lathe that was being used in
the West, now became the general tool for shaping jade. A free-
standing wheel was used for reeling silk, while a large wheel fitted
with edge fans was used for winnowing grain. Two centuries later
the latter device was to be adapted to result in an even more
efficient iron industry. With the introduction of the water wheel,
the winnowing fan was converted into a blow engine that gave an
even more regular draught through the furnace and fining ovens,
while the triphammer, instead of being worked by foot, was now
worked by cams on the shaft of the water wheel. By the fifth century A.D. the Chinese were producing both cast iron and wrought
iron and, by fusion of the two metals, were able to provide steel of

good quality. This industry was, hence, far in advance of that in the West.

Space does not allow us to elaborate upon the many other inventions of this epoch in China, but the first five centuries A.D. also saw the reintroduction of stoneware, now in a far more refined form, and the invention of paper. The latter was made from a mash of plant fibers which was spread evenly on the bottom of a fine sieve, allowed to drain, and the thin sheet so produced then stuck against a wall in the full sun to dry. At first sight this would appear to be a totally Chinese development, but the fact that the nomads had long since been using a felt made of animal fibers should not be overlooked, for in essence paper is merely a felt made of vegetable matter rather than of animal hair.

Throughout this all too brief survey of the early development of Chinese technology we have frequently referred to the nomads, as we did when speaking of the main stream of technological development in western Asia. It is now to these people that we must turn.

THE NOMADS

AT THE VERY BEGINNING of this book the reader was warned that the history of technology in these early periods would, through lack of knowledge, have to be incomplete. We now come to one of the largest gaps in our history, for what can honestly be said about the people of the steppes is remarkably little. The reason for this lack of information is simple. They were a people largely without cities and most of our knowledge of them comes from the excavation of graves. Many of these were marked by a tumulus raised over the burial place, and these have, of course, attracted the attention of excavators. But many were not marked in any way and their discovery has been purely a matter of chance. Furthermore, such cities as existed in all probability occupied precisely the same sites as those of modern cities. There is, for example, no knowing how far back such a town as Samarkand could be traced in history were it possible to excavate under the present buildings. As things stand, however, we must make do with what little material is available and be forewarned that most of what follows will be conjectural.

We should not, however, allow ourselves to be confused by the word "nomad." Today we often use it to describe someone who is footloose, or we think of gypsies or other relic populations who have maintained a mobile way of life and no longer truly fit into modern society. The word is tainted and has the implication of a people who are backward or even dishonest. In the period of which we are speaking, however, there is no good reason to believe that the nomad was in most respects less advanced technologically than his sedentary neighbors; whereas the fact that he followed his herds and has left behind no obvious traces of towns or cities should not allow us to think of him as someone substandard. From the very moment that wheeled vehicles came into being the covered wagon and the tent became the nomad's house, and the few examples of such vehicles that have come to light as a result of excavation demonstrate that they were no better and no worse equipped than the mud-brick dwellings of the average peasant in Egypt or Mesopotamia. They were, if anything, probably rather more hygienic. Essentially the nomad encampment was itself a city with many of the appurtenances—both social and material—to be found in the more settled communities of this epoch.

Nevertheless the way of life imposed upon the nomads by their need to be on the move in search of fresh pastures for their cattle meant that in certain fields technological advance was bound to be inhibited. Evidently we would not expect, nor do we find, complex building techniques among these people, and by the same token, since they were not forest dwellers they were not involved in the creation of large wooden structures either in the form of buildings or of mechanical devices. Also we would not expect the nomads to be responsible for the production of either sophisticated ceramics or of glassware. Nevertheless, when we attempt to assess the role of the nomads in the development of other technologies such as metalworking or the evolution of the wheeled vehicle, the situation is totally different: our problem is chiefly to assess whether the nomads were responsible for certain innovations or merely the borrowers and transmitters of ideas.

As we have already seen, the carts and wagons as first known in the Near East were already fairly advanced. The earliest wagons which we know of on the Asiatic steppe appear to have been of comparable date, and so far as our knowledge of these things goes it would seem more than likely that vehicles with solid wooden wheels and drawn by oxen in paired draught were in fact an invention of the nomads themselves, an invention later borrowed by the people of Mesopotamia. At a later date the taming and breeding of horses was certainly a development carried out on the steppe, in which case the chariot, too, was in all probability first evolved in this area. At a still later date, the use of the horse for riding, the

FIG. 259:
Reconstruction of a steppe wagon of about 2500 B.C.,
based on surviving wooden fragments from burials and
on clay models.

The home of the Asiatic nomad was either the
ox-drawn wagon or the tent, both of which were
commonly roofed with felt. The design of neither
appears to have changed greatly over the millennia,
apart from the use of spoked wheels.

gradual development of the saddle and finally of stirrups were all unquestionably innovations for which the nomads were totally responsible.

Again, as we have already seen, the composite bow must be regarded as an invention that went hand in hand with the growth of chariots and cavalry, and there can be little doubt that the higher civilizations of the Near East and of China both acquired this weapon from the same source. Since it would be difficult to build chariots of any quality or to make sophisticated weapons such as the composite bow without metal tools, it must be assumed that the nomads became as advanced in their metalworking techniques as their contemporaries in Mesopotamia, while there can be little reasonable doubt that it was through their contacts with the nomads that the Chinese acquired a knowledge of bronzeworking, even if they themselves applied it to the making of ritual vessels rather than to the manufacture of tools and weapons. As we have already seen, early metal technology was in no way incompatible with the nomadic way of life, and there is nothing outrageous in supposing that within each nomadic society was to be found a number of bronzesmiths. Indeed, our limited knowledge of bronze tools from this vast area shows that this was the case.

What is less certain, however, is the role played by the nomads in the transmitting of iron technology, or whether indeed they may not themselves have been responsible for the ultimate developments that made it possible to use iron on a large scale. The important factor in the evolution of iron technology was, of course, the introduction of the bellows; it is of interest to notice that this device in its improved form appeared both in China and in the Near East at much the same time. One is bound to ask, therefore, whether the nomads themselves had already perfected this instrument before it was known in either of the main areas of sedentary civilization. At the risk of being accused of romanticism, it may be pointed out that Anarcharis, who reputedly invented the bellows in Ionia, was himself a Scythian and would, had such a device already existed, have been well aware of it in his own home land. Furthermore, of all the peoples of antiquity most likely to have developed this device, the nomads, in their long tradition

FIG. 260: *A silver-appliqué ornament from Ordos; about 300 B.C.*

Because of their way of life the Asiatic nomads
were unable to practice those crafts requiring
permanent workshops. In other fields, such as
metallurgy, they excelled, developing their own
distinctive forms of weapons, tools and ornamentation,
of which this seated mule may be seen as a typical
example.

of working with hides and leather, would seem to be the most
probable candidates.

Unfortunately, this argument can immediately be countered
by considering the development of the crossbow, which again
appeared at much the same time both in the Far East and the Near
East, yet which in China was apparently developed principally to
withstand the onslaught of the nomads themselves. It seems highly
unlikely that, in this case, the nomads were the agents for transmitting the invention or that they were responsible in the first
place for the evolution of the crossbow itself.

Essentially the problem of assessing the contribution made by
the nomads to the development of technology in the ancient world
lies in the fact that one can never be quite certain whether ideas
were transmitted by the nomads or whether new developments
which were similar in character, but took place at great distances
apart, were essentially independent inventions. Very much more
will have to be known about the early history of the nomads before
a realistic appraisal of their contribution to technological advance
during this period can be made.

THE NEW WORLD

IT IS THE AUTHOR'S VIEW that, correctly speaking, the development of technology in the New World, since in all probability it was totally unrelated to that of the Old World during the period under review, should have no place in this book. A very summary survey of a few aspects of technological evolution in the New World is included solely to illustrate the point just made. The difficulty of accepting the independent invention of a number of techniques may give rise to elaborate theories of past social contact to account for their appearance in widely dispersed areas. Briefly, it has been supposed by a number of writers that the early development of metal technology in the New World was the outcome of contacts made across the Pacific. The purpose of this survey is to show that this was not necessarily the case.

The nub of much technological innovation in the New World appears to have been one of the least propitious areas of all—the coastal belt of Peru and Ecuador. This narrow strip of land, varying in breadth from twenty to forty miles, is largely desert, punctuated at intervals by rivers whose sources lie high in the Andes. Ultimately each river valley was to become virtually a separate state connected to its neighbors only by limited routes across the desert. Of the early development of this region we know very little indeed, but by 1000 B.C. the valleys were clearly occupied by settled agricultural communities. Like the people of the Near East, they were building both in stone and in mud brick. They had cultivated the potato and maize as staple crops, but the wild fauna was less satisfactory for domestication than that of the Old World. Nevertheless, the llama, the vicuña and the alpaca were all being exploited, either as pack animals or to provide wool, and the guinea pig had been domesticated as a source of food. Looms, and indeed weaving techniques in general, had become highly developed, and the people of this area were producing some of the most remarkable textiles of antiquity.

Within the river valleys it was possible to grow crops by means of irrigation, but higher up in the foothills of the Andes agriculture was only possible by terracing, and it is questionable whether this was practiced at such an early date. The Andes themselves, however, were a source of prolific mineral wealth, with deposits of gold, silver, copper and tin. It was to this mountainous area that the people in the river valleys had to look for their metals.

By 800 B.C. the people of the Peruvian coastal belt had already begun to work in gold, and as far as one can judge this metal was being acquired in exactly the same way as it was first won in the Near East, namely, by the panning of placer deposits. During the subsequent five centuries there developed a school of craftsmanship devoted entirely to the hammering of gold, in which the metal was beaten into shape and in which decoration was applied by the use of punches. This so-called Chavin goldwork is of interest chiefly because it relied to a degree upon another technology, the working of stone. The early Peruvians had already become adept at shaping hard rocks by polishing and other abrasive techniques, and very frequently gold objects were made by hammering the metal into molds made of stone. The quality of craftsmanship was truly remarkable, but for five centuries there appears to have been no technological advance whatsoever, and it would appear that the metal was seldom heated, at least not sufficiently to make it melt. The end of the fourth century B.C. in Peru saw a political upheaval which as yet our limited knowledge makes it impossible to define, but in all probability the Chavin settlements were overrun by newcomers from some other area. Whatever the cause of change, how-

FIG. 261: *A gold bracelet from Peru; about 500* B.C.

FIG. 262: *Peruvian metal founding, part of a drawing from a sixteenth-century codex.*

The earliest metalwork in the New World was shaped solely by hammering. Gold ornaments, such as this bracelet, first appeared in Peru about 500 B.C. From about 300 B.C. onward, metals—chiefly gold, silver and copper—were being melted, cast, soldered and extracted from their ores. Curiously, Peruvian craftsmen found the same solution to raising the temperature of their furnaces as did metallurgists of the Near East more than two millennia previously (see Figure 46). The Peruvians were still using blowpipes at the time of the Spanish Conquest. Tapping the molten metal directly from the furnace to the mold, as shown in this picture, was not universally practiced, for crucibles of this and earlier periods are known.

ever, the effect was dramatic, bringing with it enormous advances both in the making of pottery and in metalworking.

It would appear that the new technological developments were confined to a number of nuclei each in a distinct river valley or in a number of associated river valleys, but the differences between the products of one region and another appear to have been due largely to artistic preferences rather than to the use of different techniques. Ultimately the more northerly group of people, the Mochica, were to gain political dominance over the whole area.

Hitherto, pottery, which in itself may be seen as an independent invention of the New World, was made by the simple techniques that one would associate with primitive societies anywhere in the world. At first it had been made by modeling by hand, and later by placing a ball of clay on a saucer that could be made to rotate fairly rapidly. Vessels were produced in this way that were not dissimilar from those made on a turntable in the Old World. With the appearance of the Mochica and other contemporary peoples on the Peruvian coast, however, the better-quality pottery was made almost entirely by a sophisticated casting technique. The component parts of each vessel were shaped by forcing slabs of clay into open molds, the various pieces being finally luted together to form the whole pot. Apart from the elaborately decorated wares themselves, even the handles and spouts were made

FIG. 263: *A Peruvian gold beaker; about* A.D. *1000.*

FIG. 264: *Diagram showing the stages by which a vessel of this type was made, based largely on detailed metallographic studies.*

As in the Near East, despite improved casting and soldering techniques, a large part of Peruvian metal output continued to depend upon hammerwork, thus standing in strong contrast to contemporary metal-shaping methods in China, from which quarter some authors see the inspiration for early Peruvian metallurgy.

FIG. 265: *Diagram showing how Peruvian pottery was made in a mold from about 300 B.C. onward.*

Pottery of high quality began to be made in Peru just before the beginning of our era. Slabs of clay were squeezed into opposing mold pieces which were then joined to form the complete vessel. The technique, again, stands in total contrast to contemporary practice in China, and was almost without question a local development.

by this method. Sadly, however, although the Mochica pottery is a prolific source of information about their way of life no single vessel appears to give any hint as to how they set about working metals. As we have already noted, however, in the Old World stone molds in which gold could be cast, and later ceramic molds were partly to pave the way to new metallurgical techniques.

Within the last few centuries B.C. in Peru there were to appear in rapid succession the melting of gold, its casting, its alloying with silver and later with copper, the development of hard solder, the use of hammer welding and ultimately casting by means of the lost-wax method. During much the same period in Ecuador platinum and gold were being alloyed by heating together finely divided particles of the two metals—a technique that was later to be forgotten and not repeated until the nineteenth century. It was from this area that metalworking techniques were ultimately to spread to other parts of the New World. The diffusion, however, was slow and no metalworking, for example, was known in Mexico until the tenth century A.D.

To account for the rapid development of metalworking among the Mochica and in Ecuador some writers have suggested contacts across the Pacific from various quarters, such as Japan, China and Indo-China. The view that such contacts existed is supported by a detailed examination of decorative motifs employed in either region, and indeed if one looks only at the decorative motifs the argument seems highly convincing. Broadly speaking, the hypothesis is that in the period immediately prior to 300 B.C. a number of Chinese merchants made contact with the Peruvian coast in search of gold. After 300 B.C., due to political events in China, this supposed trade drifted to southern China and Indo-China. During much the same period, theoretically ships from Japan were trading with Ecuador.

Curiously, however, the techniques being used in the New World at this period were so utterly different from those current in the Old World that one must have serious reservations about the possibility of such contact. Thus, for example, while Chinese ceramics of this period were invariably thrown on a wheel, those in Peru were being made, as we have just seen, by elaborate casting techniques. And while in China metal work was being produced exclusively by casting, in Peru it was still largely being shaped by hammering. One might suppose that were the contacts sufficiently strong to allow the people of Peru to adopt Chinese and Indo-Chinese decorative motifs they might, too, have acquired something of Chinese technology. Briefly, the answer to the apparently rapid development of metal technology among the Mochica probably lies far more closely at hand. Were one to know more about the people of the highlands of Peru and their development during these centuries, as well as the nature of the political turmoil that ended so abruptly the earlier Chavin settlements, the problem would in all probability not be one of serious debate. The Chavin period was one in which there was apparently great social stability but little technological advance. There followed a period of political chaos out of which, in a matter of a few centuries, grew a new society bursting with technological innovations. The brief survey of events in the Old World which follows will show a social situation with which we should by now be perfectly familiar.

Retrospect

W E ARE NOW in a position to speculate upon some of the causes that led to technological evolution in the ancient world. The physical requirements were clear and obvious. An abundance of suitable raw materials and adequate communications were vital essentials. Areas lacking one or the other were not likely to be centers of technological advance. One might even postulate that the greater the variety of raw materials and the greater the ease of transportation so much the greater would in theory have been the rate of technological development. But the whole story was clearly not as simple as that.

Nowhere was the rate of technological advance a steady, even, upward climb. Always one seems to be confronted with sudden bursts of technological innovation followed by long periods of virtual stagnation, to be succeeded by yet another burst followed again by a long dormant period. In many places society has climbed to a particular level of technological advance and stagnated in this position even to this day. The Neolithic societies of modern New Guinea and the Amazon are classic examples of this lack of advance, a failure which can be attributed entirely to a lack of suitable raw materials and to poor communications. But there were many other societies which, having climbed to a far higher technological level, then stagnated despite an abundance of raw materials and even excellent communications.

In effect, one of the limitations to technological evolution in antiquity seems to have been imposed by man himself, for certain social conditions were apparently inimical to further technological innovation. Indeed, authoritarian governments aiming at stable social conditions appear to have been those under which there was least technological advance. The precise reason for such a state of affairs may have varied from one society to another. It may have been that the civil servants exercised far too rigid a control or did so in an unintelligent manner, while in modern terms there may have been too much capital investment in certain technologies to allow room for drastic innovations. Whatever the cause, however, the effect was the same, being tantamount to denying the benefits of good communications and resulting in a maintenance of the status quo.

The periods of sudden technological evolution were to be seen in their most dramatic form at a time shortly after one of the stable societies had suffered a setback at the hands of one of its less technologically advanced neighbors, and provided the kernel of the older society was left more or less intact, it was invariably the intrusive population which showed the way to the new technologies. Because they were aware of their lower technological standing they were willing to learn and because, being intrusive people, they had to be politically more adaptable, it was they and not the more stable civilizations that made the progress—they were able to stand aside and look at what they saw with a critical eye, selecting from their neighbors' technologies only what they wanted. Furthermore, such people brought with them technologies of their own and by cross-fertilization—applying techniques from one to another—they were able to introduce new concepts.

If the history of technology in the ancient world is to teach us anything at all, it is to tell us that no technology can stand long on its own in glorious isolation. For a technology to thrive and to develop required the presence of other often quite unrelated technologies alongside it from which ideas could be loaned and borrowed. Such was the case, anyway, in the ancient world. One is left wondering whether the world of today is so very different.

Bibliographical Note

THE TWO MAIN SOURCES on which I have drawn during the writing of this book have been *A History of Technology,* Volumes I and II, edited by Charles Singer, E. J. Holmyard, and A. R. Hall (Oxford; 1954 and 1956); and *Studies in Ancient Technology* by R. J. Forbes (Leiden; 1955 onwards), of which so far nine volumes have been published. Between them these works list all the major references to the various topics up to the dates of their publication.

On the subject of the geographical background to the development of early technology the reader will probably find *The Geography Behind History* by Gordon East (London; 1939) the most useful introduction; while the problems posed by Greek and Roman technology are very fully discussed in *Greek Science* by Benjamin Farrington (London; 1961).

In the period since *A History of Technology* and *Studies in Ancient Technology* first appeared, a number of books have been published that add to and in some cases must make us modify our views on many subjects. Among these the *Dictionnaire archéologique des techniques,* Éditions de l'Accueil (Paris; 1963), is useful since it covers China and the New World, areas barely touched upon in the former volumes.

The mechanics of mankind's evolution from hunter to settled agriculturalist has been brought under closer scrutiny than ever before in *The Domestication and Exploitation of Plants and Animals,* edited by P. J. Ucko and G. W. Dimbleby (London; 1969); while the social changes that went hand-in-glove with developing technology are considered at length in *The Primitive*

World and Its Transformations by Robert Redfield (London; 1968). The machinery of change, both social and technological, is discussed fully in *Innovation: The Basis of Cultural Change* by H. G. Barnett (London and New York; 1953).

On a more limited front, *Chemistry and Chemical Technology in Ancient Mesopotamia* by M. Levey (Amsterdam; 1954) gives a tantalizing taste of the storehouse of information still locked up in the as-yet-unread Mesopotamian cuneiform texts.

The subject of early shipping in the Mediterranean is surveyed in its entirety in *The Ancient Mariners* by L. Casson (London; 1959) and in *Marines antiques de la Méditerranée* by Jean Meirat (Paris; 1964); while *Oared Fighting Ships* by R. C. Anderson (London; 1962) and *Greek Oared Ships* by J. S. Morrison and R. T. Williams (Cambridge, Eng.; 1968) spell out in detail all that is really known about Greek and Roman oared craft. At the same time, our understanding of the nature of early sea trade in the Mediterranean has been very greatly amplified by the detailed report of an underwater excavation in "Cape Gelidonya: A Bronze Age Shipwreck" by George F. Bass in *Transactions of the American Philosophical Society*, 57 (1967), Part 8.

The evolution of early wheeled vehicles has been discussed fully in *Die Landfahrzeuge des Alten Mesopotamien* by A. Salonen (Bonn; 1951); while fresh material from Central Asia is put forward in "The Earliest Wheeled Vehicles and the Caucasian Evidence" by Stuart Piggott in the *Proceedings of the Prehistoric Society* (London), 34 (1968), pp. 266–318.

Many studies dealing with metalworking and ceramics in antiquity have appeared in the last decade. Of these, *A History of Metals* by L. Aitcheson (London; 1960) gives, perhaps, the clearest account to date of the development of early metallurgy. *The Techniques of the Attic Potter* by J. V. Noble (London and New York; 1966) sets out in detail the changes that took place in the middle of the first millennium B.C. that were to dominate completely Greek and Roman ceramic technology.

The rapid development of iron technology in China during the critical centuries following the middle of the first millennium

B.C. is fully discussed in *The Development of Iron and Steel Technology in China* by J. Needham (London; 1958); while all the major aspects of early Chinese technology are dealt with in *Archaeology in China* (four volumes) by Chêng Tê-K'un (Cambridge, Eng.; 1959 onwards).

The development of early metallurgy in the New World is the subject of *Sweat of the Sun and Tears of the Moon* by André Emmerich (Washington; 1965). A shorter, but very useful account of the same subject matter is to be found in "Early Metallurgy in the New World" by D. T. Easby in *Scientific American* (April 1966).

Index

Numbers in italics refer to pages on which illustrations can be found

A NOTE ABOUT THE AUTHOR

Henry Hodges was born in 1920 at Deddington, Oxford, England, and was educated at Kimbolton School; St. John's College, Oxford; and London University. He served in the Royal Navy during World War II, and has been a master of Heatherdown School, Ascot, and Assistant Lecturer for Queen's University, Belfast. He has been on the faculty of London University since 1957 and is now Senior Lecturer in Archaeology. Since 1964 he has been a Fellow of the International Institute for Conservation. Mr. Hodges is a member of the editorial board of *World Archaeology* and the author of *Artifacts: An Introduction to Early Materials and Technology,* which was published in this country as well as in England in 1964. He is married and has two children.

A NOTE ON THE TYPE

The text of this book is set in Caledonia, a Linotype face designed by W. A. Dwiggins. It belongs to the family of printing types called "modern face" by printers—a term used to mark the change in style of type letters that occurred about 1800. Caledonia borders on the general design of Scotch Modern, but is more freely drawn than that letter.

The book was composed and bound by The Book Press, Brattleboro, Vermont, and printed by The Halliday Lithograph Corporation, West Hanover, Massachusetts.

The typography and binding design are by Anita Karl.